Pitman Research Notes in Mathematics Series

Submission of proposals for consideration
Suggestions for publication, in the form of outlines and representative samples, are invited by the Editorial Board for assessment. Intending authors should approach one of the main editors or another member of the Editorial Board, citing the relevant AMS subject classifications. Alternatively, outlines may be sent directly to the publisher's offices. Refereeing is by members of the board and other mathematical authorities in the topic concerned, throughout the world.

Preparation of accepted manuscripts
On acceptance of a proposal, the publisher will supply full instructions for the preparation of manuscripts in a form suitable for direct photo-lithographic reproduction. Specially printed grid sheets are provided and a contribution is offered by the publisher towards the cost of typing. Word processor output, subject to the publisher's approval, is also acceptable.

Illustrations should be prepared by the authors, ready for direct reproduction without further improvement. The use of hand-drawn symbols should be avoided wherever possible, in order to maintain maximum clarity of the text.

The publisher will be pleased to give any guidance necessary during the preparation of a typescript, and will be happy to answer any queries.

Important note
In order to avoid later retyping, intending authors are strongly urged not to begin final preparation of a typescript before receiving the publisher's guidelines and special paper. In this way it is hoped to preserve the uniform appearance of the series.

Longman Scientific & Technical
Longman House
Burnt Mill
Harlow, Essex, UK
(tel (0279) 26721)

Titles in this series

1 Improperly posed boundary value problems
A Carasso and A P Stone
2 Lie algebras generated by finite dimensional ideals
I N Stewart
3 Bifurcation problems in nonlinear elasticity
R W Dickey
4 Partial differential equations in the complex domain
D L Colton
5 Quasilinear hyperbolic systems and waves
A Jeffrey
6 Solution of boundary value problems by the method of integral operators
D L Colton
7 Taylor expansions and catastrophes
T Poston and I N Stewart
8 Function theoretic methods in differential equations
R P Gilbert and R J Weinacht
9 Differential topology with a view to applications
D R J Chillingworth
10 Characteristic classes of foliations
H V Pittie
11 Stochastic integration and generalized martingales
A U Kussmaul
12 Zeta-functions: An introduction to algebraic geometry
A D Thomas
13 Explicit *a priori* inequalities with applications to boundary value problems
V G Sigillito
14 Nonlinear diffusion
W E Fitzgibbon III and H F Walker
15 Unsolved problems concerning lattice points
J Hammer
16 Edge-colourings of graphs
S Fiorini and R J Wilson
17 Nonlinear analysis and mechanics: Heriot-Watt Symposium Volume I
R J Knops
18 Actions of fine abelian groups
C Kosniowski
19 Closed graph theorems and webbed spaces
M De Wilde
20 Singular perturbation techniques applied to integro-differential equations
H Grabmüller
21 Retarded functional differential equations: A global point of view
S E A Mohammed
22 Multiparameter spectral theory in Hilbert space
B D Sleeman
24 Mathematical modelling techniques
R Aris
25 Singular points of smooth mappings
C G Gibson
26 Nonlinear evolution equations solvable by the spectral transform
F Calogero
27 Nonlinear analysis and mechanics: Heriot-Watt Symposium Volume II
R J Knops
28 Constructive functional analysis
D S Bridges
29 Elongational flows: Aspects of the behaviour of model elasticoviscous fluids
C J S Petrie
30 Nonlinear analysis and mechanics: Heriot-Watt Symposium Volume III
R J Knops
31 Fractional calculus and integral transforms of generalized functions
A C McBride
32 Complex manifold techniques in theoretical physics
D E Lerner and P D Sommers
33 Hilbert's third problem: scissors congruence
C-H Sah
34 Graph theory and combinatorics
R J Wilson
35 The Tricomi equation with applications to the theory of plane transonic flow
A R Manwell
36 Abstract differential equations
S D Zaidman
37 Advances in twistor theory
L P Hughston and R S Ward
38 Operator theory and functional analysis
I Erdelyi
39 Nonlinear analysis and mechanics: Heriot-Watt Symposium Volume IV
R J Knops
40 Singular systems of differential equations
S L Campbell
41 N-dimensional crystallography
R L E Schwarzenberger
42 Nonlinear partial differential equations in physical problems
D Graffi
43 Shifts and periodicity for right invertible operators
D Przeworska-Rolewicz
44 Rings with chain conditions
A W Chatters and C R Hajarnavis
45 Moduli, deformations and classifications of compact complex manifolds
D Sundararaman
46 Nonlinear problems of analysis in geometry and mechanics
M Atteia, D Bancel and I Gumowski
47 Algorithmic methods in optimal control
W A Gruver and E Sachs
48 Abstract Cauchy problems and functional differential equations
F Kappel and W Schappacher
49 Sequence spaces
W H Ruckle
50 Recent contributions to nonlinear partial differential equations
H Berestycki and H Brezis
51 Subnormal operators
J B Conway

52 Wave propagation in viscoelastic media
F Mainardi
53 Nonlinear partial differential equations and their applications: Collège de France Seminar. Volume I
H Brezis and J L Lions
54 Geometry of Coxeter groups
H Hiller
55 Cusps of Gauss mappings
T Banchoff, T Gaffney and C McCrory
56 An approach to algebraic K-theory
A J Berrick
57 Convex analysis and optimization
J-P Aubin and R B Vintner
58 Convex analysis with applications in the differentiation of convex functions
J R Giles
59 Weak and variational methods for moving boundary problems
C M Elliott and J R Ockendon
60 Nonlinear partial differential equations and their applications: Collège de France. Seminar. Volume II
H Brezis and J L Lions
61 Singular systems of differential equations II
S L Campbell
62 Rates of convergence in the central limit theorem
Peter Hall
63 Solution of differential equations by means of one-parameter groups
J M Hill
64 Hankel operators on Hilbert space
S C Power
65 Schrödinger-type operators with continuous spectra
M S P Eastham and H Kalf
66 Recent applications of generalized inverses
S L Campbell
67 Riesz and Fredholm theory in Banach algebra
B A Barnes, G J Murphy, M R F Smyth and T T West
68 Evolution equations and their applications
F Kappel and W Schappacher
69 Generalized solutions of Hamilton-Jacobi equations
P L Lions
70 Nonlinear partial differential equations and their applications: Collège de France Seminar. Volume III
H Brezis and J L Lions
71 Spectral theory and wave operators for the Schrödinger equation
A M Berthier
72 Approximation of Hilbert space operators I
D A Herrero
73 Vector valued Nevanlinna Theory
H J W Ziegler
74 Instability, nonexistence and weighted energy methods in fluid dynamics and related theories
B Straughan
75 Local bifurcation and symmetry
A Vanderbauwhede
76 Clifford analysis
F Brackx, R Delanghe and F Sommen
77 Nonlinear equivalence, reduction of PDEs to ODEs and fast convergent numerical methods
E E Rosinger
78 Free boundary problems, theory and applications. Volume I
A Fasano and M Primicerio
79 Free boundary problems, theory and applications. Volume II
A Fasano and M Primicerio
80 Symplectic geometry
A Crumeyrolle and J Grifone
81 An algorithmic analysis of a communication model with retransmission of flawed messages
D M Lucantoni
82 Geometric games and their applications
W H Ruckle
83 Additive groups of rings
S Feigelstock
84 Nonlinear partial differential equations and their applications: Collège de France Seminar. Volume IV
H Brezis and J L Lions
85 Multiplicative functionals on topological algebras
T Husain
86 Hamilton-Jacobi equations in Hilbert spaces
V Barbu and G Da Prato
87 Harmonic maps with symmetry, harmonic morphisms and deformations of metrics
P Baird
88 Similarity solutions of nonlinear partial differential equations
L Dresner
89 Contributions to nonlinear partial differential equations
C Bardos, A Damlamian, J I Díaz and J Hernández
90 Banach and Hilbert spaces of vector-valued functions
J Burbea and P Masani
91 Control and observation of neutral systems
D Salamon
92 Banach bundles, Banach modules and automorphisms of C*-algebras
M J Dupré and R M Gillette
93 Nonlinear partial differential equations and their applications: Collège de France Seminar. Volume V
H Brezis and J L Lions
94 Computer algebra in applied mathematics: an introduction to MACSYMA
R H Rand
95 Advances in nonlinear waves. Volume I
L Debnath
96 FC-groups
M J Tomkinson
97 Topics in relaxation and ellipsoidal methods
M Akgül
98 Analogue of the group algebra for topological semigroups
H Dzinotyiweyi
99 Stochastic functional differential equations
S E A Mohammed

100 Optimal control of variational inequalities
V Barbu
101 Partial differential equations and dynamical systems
W E Fitzgibbon III
102 Approximation of Hilbert space operators. Volume II
C Apostol, L A Fialkow, D A Herrero and D Voiculescu
103 Nondiscrete induction and iterative processes
V Ptak and F-A Potra
104 Analytic functions – growth aspects
O P Juneja and G P Kapoor
105 Theory of Tikhonov regularization for Fredholm equations of the first kind
C W Groetsch
106 Nonlinear partial differential equations and free boundaries. Volume I
J I Díaz
107 Tight and taut immersions of manifolds
T E Cecil and P J Ryan
108 A layering method for viscous, incompressible L_p flows occupying R^n
A Douglis and E B Fabes
109 Nonlinear partial differential equations and their applications: Collège de France Seminar. Volume VI
H Brezis and J L Lions
110 Finite generalized quadrangles
S E Payne and J A Thas
111 Advances in nonlinear waves. Volume II
L Debnath
112 Topics in several complex variables
E Ramírez de Arellano and D Sundararaman
113 Differential equations, flow invariance and applications
N H Pavel
114 Geometrical combinatorics
F C Holroyd and R J Wilson
115 Generators of strongly continuous semigroups
J A van Casteren
116 Growth of algebras and Gelfand–Kirillov dimension
G R Krause and T H Lenagan
117 Theory of bases and cones
P K Kamthan and M Gupta
118 Linear groups and permutations
A R Camina and E A Whelan
119 General Wiener–Hopf factorization methods
F-O Speck
120 Free boundary problems: applications and theory, Volume III
A Bossavit, A Damlamian and M Fremond
121 Free boundary problems: applications and theory, Volume IV
A Bossavit, A Damlamian and M Fremond
122 Nonlinear partial differential equations and their applications: Collège de France Seminar. Volume VII
H Brezis and J L Lions
123 Geometric methods in operator algebras
H Araki and E G Effros
124 Infinite dimensional analysis–stochastic processes
S Albeverio
125 Ennio de Giorgi Colloquium
P Krée
126 Almost-periodic functions in abstract spaces
S Zaidman
127 Nonlinear variational problems
A Marino, L Modica, S Spagnolo and M Degiovanni
128 Second-order systems of partial differential equations in the plane
L K Hua, W Lin and C-Q Wu
129 Asymptotics of high-order ordinary differential equations
R B Paris and A D Wood
130 Stochastic differential equations
R Wu
131 Differential geometry
L A Cordero
132 Nonlinear differential equations
J K Hale and P Martinez-Amores
133 Approximation theory and applications
S P Singh
134 Near-rings and their links with groups
J D P Meldrum
135 Estimating eigenvalues with *a posteriori/a prior* inequalities
J R Kuttler and V G Sigillito
136 Regular semigroups as extensions
F J Pastijn and M Petrich
137 Representations of rank one Lie groups
D H Collingwood
138 Fractional calculus
G F Roach and A C McBride
139 Hamilton's principle in continuum mechanics
A Bedford
140 Numerical analysis
D F Griffiths and G A Watson
141 Semigroups, theory and applications. Volume
H Brezis, M G Crandall and F Kappel
142 Distribution theorems of L-functions
D Joyner
143 Recent developments in structured continua
D De Kee and P Kaloni
144 Functional analysis and two-point differential operators
J Locker
145 Numerical methods for partial differential equations
S I Hariharan and T H Moulden
146 Completely bounded maps and dilations
V I Paulsen
147 Harmonic analysis on the Heisenberg nilpoten Lie group
W Schempp
148 Contributions to modern calculus of variations
L Cesari
149 Nonlinear parabolic equations: qualitative properties of solutions
L Boccardo and A Tesei
150 From local times to global geometry, control a physics
K D Elworthy

151 A stochastic maximum principle for optimal control of diffusions
U G Haussmann
152 Semigroups, theory and applications. Volume II
H Brezis, M G Crandall and F Kappel
153 A general theory of integration in function spaces
P Muldowney
154 Oakland Conference on partial differential equations and applied mathematics
L R Bragg and J W Dettman
155 Contributions to nonlinear partial differential equations. Volume II
J I Díaz and P L Lions
156 Semigroups of linear operators: an introduction
A C McBride
157 Ordinary and partial differential equations
B D Sleeman and R J Jarvis
158 Hyperbolic equations
F Colombini and M K V Murthy
159 Linear topologies on a ring: an overview
J S Golan
160 Dynamical systems and bifurcation theory
M I Camacho, M J Pacifico and F Takens
161 Branched coverings and algebraic functions
M Namba
162 Perturbation bounds for matrix eigenvalues
R Bhatia
163 Defect minimization in operator equations: theory and applications
R Reemtsen
164 Multidimensional Brownian excursions and potential theory
K Burdzy
165 Viscosity solutions and optimal control
R J Elliott
166 Nonlinear partial differential equations and their applications. Collège de France Seminar. Volume VIII
H Brezis and J L Lions
167 Theory and applications of inverse problems
H Haario
168 Energy stability and convection
G P Galdi and B Straughan
169 Additive groups of rings. Volume II
S Feigelstock
170 Numerical analysis 1987
D F Griffiths and G A Watson
171 Surveys of some recent results in operator theory. Volume I
J B Conway and B B Morrel
172 Amenable Banach algebras
J-P Pier
173 Pseudo-orbits of contact forms
A Bahri
174 Poisson algebras and Poisson manifolds
K H Bhaskara and K Viswanath
175 Maximum principles and eigenvalue problems in partial differential equations
P W Schaefer
176 Mathematical analysis of nonlinear, dynamic processes
K U Grusa
177 Cordes' two-parameter spectral representation theory
D F McGhee and R H Picard
178 Equivariant K-theory for proper actions
N C Phillips
179 Elliptic operators, topology and asymptotic methods
J Roe
180 Nonlinear evolution equations
J K Engelbrecht, V E Fridman and E N Pelinovski
181 Nonlinear partial differential equations and their applications. Collège de France Seminar. Volume IX
H Brezis and J L Lions
182 Critical points at infinity in some variational problems
A Bahri
183 Recent developments in hyperbolic equations
L Cattabriga, F Colombini, M K V Murthy and S Spagnolo
184 Optimization and identification of systems governed by evolution equations on Banach space
N U Ahmed
185 Free boundary problems: theory and applications. Volume I
K H Hoffmann and J Sprekels
186 Free boundary problems: theory and applications. Volume II
K H Hoffmann and J Sprekels
187 An introduction to intersection homology theory
F Kirwan
188 Derivatives, nuclei and dimensions on the frame of torsion theories
J S Golan and H Simmons
189 Theory of reproducing kernels and its applications
S Saitoh
190 Volterra integrodifferential equations in Banach spaces and applications
G Da Prato and M Iannelli
191 Nest algebras
K R Davidson
192 Surveys of some recent results in operator theory. Volume II
J B Conway and B B Morrel
193 Nonlinear variational problems. Volume II
A Marino and M K Murthy
194 Stochastic processes with multidimensional parameter
M E Dozzi
195 Prestressed bodies
D Iesan
196 Hilbert space approach to some classical transforms
R H Picard
197 Stochastic calculus in application
J R Norris

Prestressed bodies

D Ieşan

University of Iaşi, Romania

Prestressed bodies

Longman Scientific & Technical

Copublished in the United States with
John Wiley & Sons, Inc., New York

Longman Scientific & Technical
Longman Group UK Limited
Longman House, Burnt Mill, Harlow
Essex CM20 2JE, England
and Associated Companies throughout the world.

Copublished in the United States with
John Wiley & Sons, Inc., 605 Third Avenue, New York, NY 10158

First published 1989

AMS Subject Classifications: (main) 73C35, 73R05, 73J05
(subsidiary) 73K05, 35B30, 35L55

ISSN 0269-3674

British Library Cataloguing in Publication Data
Ieşan, D.
Prestressed bodies.
1. Elasticity theory
I. Title
531′.3823

ISBN 0-582-03761-1

Library of Congress Cataloging-in-Publication Data
Ieşan, Dorin.
Prestressed bodies.
(Pitman research notes in mathematics series,
0269-3674 ; 195)
1. Elasticity. 2. Continuum mechanics. I. Title.
II. Series.
QA931.I36 1988 531′.3823 88-8463
ISBN 0-470-21267-5

Printed and bound in Great Britain by
Biddles Ltd, Guildford and King's Lynn

Contents

Introduction

This work is concerned mainly with basic problems of the theory of the initially stressed elastic bodies. On the whole, the subject matter is directed towards recent developments.

The theory of initially stressed bodies is of considerable interest both from the mathematical and the technical point of view. We present a systematic treatment of the linear theory of infinitesimal elastic deformations superimposed on finite deformations. This theory provides a natural extension of the classical theory of elasticity, as the initial configuration needs no longer be unstressed. It also establishes a basis for discussing stability of finite deformations. The problem of stability, however, is not considered here. The reader interested in this subject will find a full account in the monograph of Knops and Wilkes [229]. We have attempted to isolate those conceptual and mathematical difficulties which arise over and above those inherent in the problems concerned with unstressed bodies. The greatest of these arises from the dependence of the coefficients of the field equations (obtained for the description of the superimposed motion) on the functions describing the initial deformation. These coefficients, in general, are functions of both position and time. The theory of initially stressed bodies has given impetus to theoretical research into the equations of elastic bodies for which little or no information is known concerning the elasticities (see, for example, Knops and Payne [227], Sections 2.1, 4.4, 8.3).

Chapter 1 is concerned mainly with the development of the fundamental equations of the linear theory of initially stressed elastic bodies. Chapter 2 is devoted to a study of the boundary-value problems of elastostatics. Chapter 3 deals with the boundary-initial-value problems of the dynamical theory of elastic bodies. Chapter 4 is concerned with the theory of infinitesimal thermoelastic deformations superimposed on a large one. Chapters 5 and 6 contain the dynamic and equilibrium theories of thermoelastic bodies, respectively. The basic equations and smoothness hypotheses are given in the first section of each chapter. Our discussion is restricted to general compressible solids. In Chapter 7 there are presented the basic equations for infinitesimal displacements and weak fields superimposed on finite deformations in an elastic dielectric. The photoelastic effect is only one example of numerous phenomena which are predicted by this theory.

To review the vast literature on applications and special problems is not our intention. The illustrations included are examples considered relevant to the purpose of the text. An account of the historical developments as well as references to other various contributions, may be found in the books and some of the papers cited.

Most of the material is self-contained and the only prerequisite is an elementary course in theory of elasticity. We make no claim to completeness. Neither the contents, nor the list of works cited are exhaustive. Nevertheless, it is hoped that the present work gives a unified, accessible treatment of a part of the contributions that have been made to the subject.

1 The linear theory of elasticity for initially stressed bodies

1.1 Notation and preliminary results

We consider a body that at time t_0 occupies the region B_0 of Euclidean three-dimensional space E^3. Throughout this work, unless specified to the contrary, B_0 will denote a bounded regular region (see, for example, Gurtin [153], Sect. 5). The configuration of the body at time t_0 is taken as reference configuration.

The motion of the body is referred to the reference configuration and a fixed system of rectangular Cartesian axes. We assume that B_0 is a configuration of equilibrium. The Cartesian coordinates of a typical particle in the reference configuration are X_A. The current position of the particle at time t is denoted by x_i where

$$x_i = x_i(X_1,X_2,X_3,t), \tag{1.1.1}$$

with the condition

$$\det\left(\frac{\partial x_i}{\partial X_A}\right) > 0. \tag{1.1.2}$$

We shall employ the usual summation and differentiation conventions: lower and upper case Latin subscripts (unless otherwise specified) are understood to range over the integers (1,2,3) whereas Greek subscripts are confined to the range (1,2), summation over repeated subscripts is implied and subscripts preceded by a comma denote partial differentiation with respect to the corresponding Cartesian coordinate. Throughout this work, a superposed dot denotes the material derivative with respect to the time. Letters marked by an underbar stand for tensors of an order $p \geqq 1$, and if $\underline{v}$ has the order p, we write $v_{ij\ldots k}$ (p subscripts) for the rectangular Cartesian components of $\underline{v}$.

We assume the continuous differentiability of x_i with respect to each of the variables X_A and t as many times as required.

Let (t_0,t_1) denote a given interval of time. A motion of the body is a class C^2 vector field $\underline{x}$ on $B_0 \times (t_0,t_1)$. We say that g is of class $C^{M,N}$ on $B_0 \times (t_0,t_1)$ if g is continuous on $B_0 \times (t_0,t_1)$ and the functions

$$\frac{\partial^m}{\partial X_A\, \partial X_B \cdots \partial X_K}\left(\frac{\partial^n g}{\partial t^n}\right),\ m \in \{0,1,\ldots,M\},\ n \in \{0,1,\ldots,N\},\ m + n \leq \max(M,N),$$

exist and are continuous on $B_0 \times (t_0,t_1)$. We write C^N for $C^{N,N}$.

We consider an arbitrary material volume P in the continuum, bounded by a surface ∂P at time t, and we suppose that P is the corresponding region in the reference configuration, bounded by a surface ∂P. Let the outward unit normal at ∂P be N_A, referred to the our rectangular frame of reference.

We postulate an energy balance in the form

$$\int_P \rho_0 \dot{x}_i \ddot{x}_i dV + \int_P \rho_0 \dot{\varepsilon} dV = \int_P \rho_0 f_i \dot{x}_i dV + \int_{\partial P} T_i \dot{x}_i dA, \tag{1.1.3}$$

for every part P of B_0 and every time t. Here, dV and dA denote the elements of volume and surface area respectively, ρ_0 is the density in the reference configuration, ε is the internal energy per unit mass, f_i is the body force per unit mass, and T_i is the stress vector associated with the surface ∂P but measured per unit area of the surface ∂P.

Following the procedure of Green and Rivlin [139], we consider a second motion which differs from the given motion only by a constant superposed rigid body translational velocity, the body occupying the same position at time t, and we assume that $\dot{\varepsilon}$, f_i, and T_i are unaltered by such superposed rigid velocity. If we use the equation (1.1.3) with x_i replaced by $\dot{x}_i + a_i$ where a_i are arbitrary constants, we obtain

$$\int_P \rho_0 \ddot{x}_i dV = \int_P \rho_0 f_i dV + \int_{\partial P} T_i dA. \tag{1.1.4}$$

If the components of the first Piola-Kirchhoff stress tensor (see, for example, Truesdell and Noll [356], Sect. 43 A) are T_{Ai}, it follows from (1.1.4), by the usual methods, that

$$T_i = T_{Ai}\, N_A, \tag{1.1.5}$$

and

$$T_{Ai,A} + \rho_o f_i = \rho_o \ddot{x}_i \quad \text{on} \quad B_o \times (t_o, t_1). \tag{1.1.6}$$

Use of these relations in (1.1.3) then leads to the local equation of energy balance

$$\rho_o \dot{\varepsilon} = T_{Ai} \dot{x}_{i,A}. \tag{1.1.7}$$

If we define the second Piola-Kirchhoff stress tensor T_{AB} by

$$T_{Ai} = x_{i,B} T_{AB}, \tag{1.1.8}$$

then the local balance of energy becomes

$$\rho_o \dot{\varepsilon} = T_{AB} \dot{x}_{i,j} x_{i,B} x_{j,A}. \tag{1.1.9}$$

Let us now consider a motion of the body which differs from the given motion only by a superposed uniform rigid body angular velocity, the body occupying the same position at time t, and let us assume that $\dot{\varepsilon}$ and T_{AB} are unaltered by such motion. Then, the equation (1.1.9) leads to

$$\Omega_{ij} x_{i,B} x_{j,A} T_{AB} = 0,$$

for all arbitrary constant skew symmetric tensors Ω_{ij}. This relation yields

$$T_{AB} = T_{BA}. \tag{1.1.10}$$

Let E_{AB} be the Lagrangian strain tensor. Then,

$$E_{AB} = \frac{1}{2}(x_{i,A} x_{i,B} - \delta_{AB}), \tag{1.1.11}$$

where δ_{AB} denotes Kronecker's delta. The right Cauchy-Green tensor C_{AB} and the left Cauchy-Green tensor b_{ij} are defined, respectively, by

$$C_{AB} = x_{i,A} x_{i,B}, \qquad b_{ij} = x_{i,K} x_{j,K}. \tag{1.1.12}$$

It follows from (1.1.7), (1.1.8) and (1.1.11) that the local balance of energy takes the form

$$\rho_0 \dot{\varepsilon} = T_{AB}\dot{E}_{AB}. \tag{1.1.13}$$

An elastic material is defined as one for which the following constitutive equations hold at each material point X_A and for all time t

$$\varepsilon = \varepsilon(E_{MN}), \quad T_{AB} = T_{AB}(E_{MN}). \tag{1.1.14}$$

We assume that the constitutive functions are of class C^2 on their domain, which is the set of all E_{AB}, with $E_{AB} = E_{BA}$. Clearly, the constitutive equations (1.1.14) are unaltered by superposed rigid-body motions.

The values of $\dot{E}_{KL}$ are arbitrary, subject to $\dot{E}_{KL} = \dot{E}_{LK}$, so it follows from (1.1.13) that

$$T_{AB} = \frac{\partial e}{\partial E_{AB}}, \tag{1.1.15}$$

where $e = \rho_0 \varepsilon$.

To avoid ambiguity we assume that ε in (1.1.14) is arranged as a symmetric function of E_{AB}, and E_{AB} is understood to mean $(E_{AB} + E_{BA})/2$ in $\partial e/\partial E_{AB}$. The basic equations of nonlinear elasticity consists of the equations of motion (1.1.6), the constitutive equations (1.1.15), and the geometrical equations (1.1.11).

We assume given surface displacement on $\bar{\Sigma}_1 \times [t_0, t_1)$ and surface force on $\Sigma_2 \times [t_0, t_1)$ where Σ_1 and Σ_2 are complementary regular subsurfaces of the boundary ∂B_0 of B_0. Thus, we have the following boundary conditions

$$x_i = \hat{x}_i \text{ on } \bar{\Sigma}_1 \times [t_0, t_1), \quad T_{Ai}N_A = \hat{T}_i \text{ on } \Sigma_2 \times [t_0, t_1), \tag{1.1.16}$$

where $\hat{x}_i$ and $\hat{T}_i$ are prescribed functions. We assume that $\hat{x}_i$ is continuous on $\bar{\Sigma}_1 \times [t_0, t_1)$ and $\hat{T}_i$ is continuous in time and piecewise regular on $\Sigma_2 \times [t_0, t_1)$.

In what follows it is useful to have the above relations expressed in terms of quantities referred to the current configuration B. Let t_{ij} be the Cauchy stress tensor. The equations of motion become

$$t_{ji,j} + \rho f_i = \rho \ddot{x}_i \quad \text{on} \quad B \times (t_0, t_1), \tag{1.1.17}$$

where ρ is the mass density of the material at time t. The boundary conditions corresponding to (1.1.16) are

$$x_i = \tilde{x}_i \text{ on } \bar{S}_1 \times [t_0, t_1), \; t_{ji}\, n_j = \tilde{t}_i \text{ on } S_2 \times [t_0, t_1), \tag{1.1.18}$$

where $\tilde{x}_i$ and $\tilde{t}_i$ are specified functions, and n_j is the unit outward normal on the surface ∂B. Clearly, S_α is the image of Σ_α by motion. We have $\partial B = \bar{S}_1 \cup S_2$, and $S_1 \cap S_2 = \emptyset$. In (1.1.18), $\tilde{t}_i$ is the surface traction per unit area of the deformed body.

We record the following relations between the stress tensors (see, for example, Truesdell and Noll [356], Sect. 43 A)

$$Jt_{ij} = x_{i,A} T_{Aj} = x_{i,A} x_{j,B} T_{AB}, \tag{1.1.19}$$

where

$$J = \det (x_{i,A}). \tag{1.1.20}$$

We also note that $\rho_0 = \rho J$.

1.2 Equations of perturbed motion

We consider three states of the body: the state B_0, and two current configurations, B and B*. We call B the primary state, and B* the secondary state. We shall designate as incremental those quantities associated with the difference of motions between the secondary and primary states. The quantities associated with the secondary state will be denoted with an asterisk and the position coordinates of the particle X_A at time t will be denoted by y_i. We have

$$y_i = y_i(X_1, X_2, X_3, t), \quad (X_A, t) \in B_0 \times (t_0, t_1), \tag{1.2.1}$$

with

$$J^* = \det (y_{i,A}) > 0. \tag{1.2.2}$$

Now let

$$u_i = y_i - x_i. \tag{1.2.3}$$

We assume that u_i is small, i.e.

$$u_i = \varepsilon' u_i', \tag{1.2.4}$$

where ε' is a constant which is small so that squares and higher powers of ε' may be neglected compared with ε', and u_i' is independent of ε'. We are therefore concerned with the equations of infinitesimal elastic deformations superimposed upon large.

In this section we shall establish the equations, boundary conditions, and initial conditions for u_i when the motion of B and loadings associated with B and B* are given.

The theory is a special case of one due to Cauchy [50]. Subsequent derivations of the basic equations for infinitesimal elastic displacements superimposed on a finite deformation are numerous (see, for example, Green, Rivlin and Shield [133], Green and Zerna [140], Ericksen and Toupin [112], Toupin and Bernstein [350], Truesdell and Noll [356], Knops and Wilkes [229]).

If we refer all quantities to the configuration B_o, then we have

$$u_i = u_i(X_1, X_2, X_3, t), \quad (X_A, t) \in B_o \times (t_o, t_1). \tag{1.2.5}$$

It is sometimes convenient to refer all quantities to the configuration of the body in its primary motion at time t instead of the reference configuration at time t_o. This is especially useful when the primary state B is a configuration of equilibrium. Then, we have

$$u_i = u_i(x_1, x_2, x_3, t), \quad (x_j, t) \in B \times (t_o, t_1). \tag{1.2.6}$$

In the secondary state B* we consider the following stress tensors: t^*_{ij} is the Cauchy stress tensor, T^*_{Ai} and T^*_{AB} are the Piola-Kirchhoff stress tensors measured per unit area in the configuration B_o, and $T^{*(1)}_{ki}$ is the first Piola-Kirchhoff stress tensor measured per unit area in the configuration B. Thus, we have

$$J^* t^*_{ij} = y_{i,A} T^*_{Aj} = y_{i,A} y_{j,B} T^*_{AB},$$
$$J' t^*_{ij} = y_{i,s} T^{*(1)}_{sj} \tag{1.2.7}$$

where

$$J^* = J'J, \; J' = \det (y_{i,s}). \tag{1.2.8}$$

Throughout, we use the notation $f_{,i} = \partial f/\partial x_i$. To a second order approximation, we obtain

$$J' = 1 + u_{s,s}. \tag{1.2.9}$$

The equations of motion for the secondary state may be written in the forms

$$T^*_{Ai,A} + \rho_0 f^*_i = \rho_0 \ddot{y}_i \quad \text{on } B_0 \times (t_0, t_1), \tag{1.2.10}$$

$$T^{*(1)}_{ji,j} + \rho f^*_i = \rho \ddot{y}_i \quad \text{on } B \times (t_0, t_1). \tag{1.2.11}$$

The constitutive equations lead to

$$e^* = e\,(E^*_{AB}), \; T^*_{AB} = \frac{\partial e^*}{\partial E^*_{AB}}. \tag{1.2.12}$$

Clearly,

$$2E^*_{AB} = y_{i,A} y_{i,B} - \delta_{AB}. \tag{1.2.13}$$

We consider the following boundary conditions corresponding to the secondary state

$$y_i = \hat{y}_i \text{ on } \bar{\Sigma}_1 \times [t_0, t_1), \quad T^*_{Ai} N_A = \hat{T}^*_i \text{ on } \Sigma_2 \times [t_0, t_1), \tag{1.2.14}$$

where $\hat{y}_i$ and $\hat{T}_i^*$ are specified functions. These conditions can be written in the form

$$y_i = \tilde{y}_i \text{ on } \bar{S}_1 \times [t_0, t_1), \quad T_{ji}^{*(1)} n_j = \tilde{T}_i^* \text{ on } S_2 \times [t_0, t_1). \tag{1.2.15}$$

Let us establish the boundary-initial-value problem which characterizes the functions u_i.

Since

$$y_{i,A} = x_{i,A} + u_{i,A} = x_{i,A} + u_{i,j} x_{j,A},$$

the relations (1.1.11) and (1.2.13) imply that

$$E_{AB}^* = E_{AB} + \frac{1}{2}(x_{i,B} u_{i,A} + x_{i,A} u_{i,B}), \tag{1.2.16}$$

or equivalently,

$$E_{AB}^* = E_{AB} + x_{i,A} x_{j,B} e_{ij}. \tag{1.2.17}$$

Here e_{ij} is defined by

$$2e_{ij} = u_{i,j} + u_{j,i}. \tag{1.2.18}$$

Then, to a second order approximation, we obtain

$$\frac{\partial e^*}{\partial E_{KL}^*} = \frac{\partial e}{\partial E_{KL}} + A_{KLMN}(E_{MN}^* - E_{MN}), \tag{1.2.19}$$

where

$$A_{KLMN} = \frac{\partial^2 e}{\partial E_{KL}\, \partial E_{MN}}. \tag{1.2.20}$$

Clearly,

$$A_{KLMN} = A_{LKMN} = A_{MNLK}. \tag{1.2.21}$$

It follows from (1.1.15), (1.2.12), (1.2.16), (1.2.19) and (1.2.21) that

$$T^{*}_{AB} = T_{AB} + A_{ABMN}\, x_{i,M}\, u_{i,N}. \tag{1.2.22}$$

From (1.1.15), (1.2.12), (1.2.17) and (1.2.19) we obtain

$$T^{*}_{AB} = T_{AB} + A_{ABMN} x_{i,M} x_{j,N} e_{ij}. \tag{1.2.23}$$

Since

$$T^{*}_{Ai} = y_{i,B} T^{*}_{AB} = (x_{i,B} + u_{i,B}) T^{*}_{AB},$$

it follows from (1.2.22) that

$$T^{*}_{Ai} = T_{Ai} + E_{iAjN} u_{j,N} + T_{AB} u_{i,B}, \tag{1.2.24}$$

where

$$E_{iKjN} = x_{i,B} x_{j,M} A_{BKMN}. \tag{1.2.25}$$

If we now define

$$S_{Ai} = T^{*}_{Ai} - T_{Ai}, \tag{1.2.26}$$

then (1.2.24) implies

$$S_{Ai} = E_{iAjN} u_{j,N} + T_{AB} u_{i,B}. \tag{1.2.27}$$

We now establish a relation between the stress tensors $T^{*(1)}_{ij}$ and t_{ij}. By (1.2.7),

$$J T^{*(1)}_{kj} = x_{k,A} T^{*}_{Aj} = x_{k,A} x_{r,B} T^{*}_{AB} y_{j,r}. \tag{1.2.28}$$

It follows from (1.1.19), (1.2.3), (1.2.23) and (1.2.28) that

$$T^{*(1)}_{ij} = t_{ij} + C_{ijrs} e_{rs} + t_{im} u_{j,m}, \tag{1.2.29}$$

where

$$C_{ijrs} = \frac{1}{J} x_{i,K} x_{j,L} x_{r,M} x_{s,N} A_{KLMN}. \tag{1.2.30}$$

Clearly,

$$C_{ijrs} = C_{jirs} = C_{rsij}. \tag{1.2.31}$$

If we define

$$s_{ij} = T^{*(1)}_{ij} - t_{ij}, \tag{1.2.32}$$

then (1.2.29) implies

$$s_{ij} = C_{ijrs} e_{rs} + t_{ir} u_{j,r}. \tag{1.2.33}$$

In the same manner we can establish the relation

$$t^*_{ij} = t_{ij} - t_{ij} u_{r,r} + t_{sj} u_{i,s} + t_{ir} u_{j,r} + C_{ijrs} e_{rs}.$$

These equations were first derived by Cauchy [50].

α) The equations of incremental motion referred to the configuration B_0. Forming differences of corresponding terms in the two sets of equations (1.1.6) and (1.2.10), we obtain

$$S_{Ai,A} + \rho_0 F_i = \rho_0 \ddot{u}_i \text{ on } B_0 \times (t_0, t_1), \tag{1.2.34}$$

where

$$F_i = f^*_i - f_i. \tag{1.2.35}$$

It follows from (1.2.27) that

$$S_{Ai} = D_{iAjN} u_{j,N}, \tag{1.2.36}$$

where

$$D_{iAjN} = E_{iAjN} + \delta_{ij} T_{AN}. \qquad (1.2.37)$$

The coefficients D_{iAjN} are evaluated in the primary state and, in general, are functions of both position and time. These quantities are determined explicitly by the deformation of the primary state. We note the symmetry properties

$$D_{iAjN} = D_{jNiA}. \qquad (1.2.38)$$

From (1.1.16) and (1.2.14) we obtain the boundary conditions

$$u_i = \hat{u}_i \text{ on } \bar{\Sigma}_1 \times [t_o,t_1), \; S_{Ai}N_A = \hat{P}_i \text{ on } \Sigma_2 \times [t_o,t_1), \qquad (1.2.39)$$

where

$$\hat{u}_i = \hat{y}_i - \hat{x}_i, \quad \hat{P}_i = \hat{T}^*_i - \hat{T}_i. \qquad (1.2.40)$$

To the equations (1.2.34), (1.2.36) and boundary conditions (1.2.39) we adjoin the initial conditions

$$u_i(\underline{X},0) = \hat{a}_i(\underline{X}), \quad \dot{u}_i(\underline{X},0) = \hat{b}_i(\underline{X}), \quad \underline{X} \in \bar{B}_o, \qquad (1.2.41)$$

where $\hat{a}_i$ and $\hat{b}_i$ are prescribed functions.

It is clearly immaterial which instant τ is selected as the initial one and hence we arbitrarily choose $\tau = 0$.

β) <u>The equations of incremental motion referred to the primary state.</u> It follows from (1.1.17), (1.2.3), (1.2.11) and (1.2.32) that

$$s_{ji,j} + \rho F_i = \rho \ddot{u}_i \text{ on } B \times (t_o,t_1). \qquad (1.2.42)$$

The relations (1.2.33) may be written in the form

$$s_{ji} = d_{ijrs}\, u_{r,s}, \qquad (1.2.43)$$

where

$$d_{ijrs} = C_{ijrs} + \delta_{ir}t_{js}. \tag{1.2.44}$$

The coefficients d_{ijrs} are functions of the state of strain of the body in the configuration B. Clearly,

$$d_{ijrs} = d_{rsij}. \tag{1.2.45}$$

Forming differences of corresponding terms in the two sets of boundary conditions (1.1.18) and (1.2.15), we obtain

$$u_i = \tilde{u}_i \text{ on } \bar{S}_1 \times [t_0,t_1), \quad s_{ji}n_j = \tilde{p}_i \text{ on } S_2 \times [t_0,t_1), \tag{1.2.46}$$

where $\tilde{u}_i = \tilde{y}_i - \tilde{x}_i$, $\tilde{p}_i = \tilde{T}_i^* - \tilde{t}_i$.

To complete the specification of the problem given by (1.2.42), (1.2.43) and (1.2.46) we must adjoin the initial conditions

$$u_i(\underline{x},0) = \tilde{a}_i(\underline{x}), \quad \dot{u}_i(\underline{x},0) = \tilde{b}_i(\underline{x}), \quad \underline{x} \in \bar{B}, \tag{1.2.47}$$

where $\tilde{a}_i$ and $\tilde{b}_i$ are given functions.

To close this section we add an expression for the internal energy function e* corresponding to the secondary state B*. Since all relations are accurate up to linear terms in ε' we must obtain the energy as far as quadratic terms in ε'. The complete strain tensor E^*_{AB} is given by

$$\begin{aligned} E^*_{AB} &= E_{AB} + x_{i,A}x_{j,B}e_{ij} + \frac{1}{2}u_{i,A}u_{i,B} = E_{AB} + \\ &+ \frac{1}{2}(x_{i,A}u_{j,B} + x_{j,B}u_{i,A}) + \frac{1}{2}u_{i,A}u_{i,B}. \end{aligned} \tag{1.2.48}$$

Since e* is a function of the strain tensor (1.2.48), retaining only linear and quadratic terms, we obtain

$$e^* = e + J[t_{ij}e_{ij} + \frac{1}{2}(C_{ijrs}e_{ij}e_{rs} + t_{ij}u_{r,i}u_{r,j})] =$$

$$= e + J[t_{ij}u_{i,j} + \frac{1}{2}(C_{ijrs}u_{i,j}u_{r,s} + t_{ij}u_{r,i}u_{r,j})] = \qquad (1.2.49)$$

$$= e + Jt_{ij}u_{i,j} + \frac{1}{2}Jd_{ijrs}u_{i,j}u_{r,s}.$$

The internal energy function e* has the alternative form

$$e^* = e + T_{Ai}u_{i,A} + \frac{1}{2}(E_{iKjN}u_{i,K}u_{j,N} + T_{AB}u_{i,A}u_{i,B}) =$$

$$= e + T_{Ai}u_{i,A} + \frac{1}{2}D_{iAjB}u_{i,A}u_{j,B}. \qquad (1.2.50)$$

The quantities D_{iAjB} and d_{ijrs} depend not only upon the material but also upon the initial strain. When the initial strain is absent then D_{iAjB} and d_{ijrs} reduce to the elasticities of the classical linear theory. Following Knops and Payne [227], henceforth we broaden the term "elasticities" to include these quantities. It follows from (1.1.19), (1.2.25), (1.2.30), (1.2.37) and (1.2.44) that

$$D_{iAjB} = JX_{A,r}X_{B,s}d_{irjs}. \qquad (1.2.51)$$

1.3 Isotropic materials

For an isotropic body we have

$$e = e(I_1, I_2, I_3), \qquad (1.3.1)$$

where I_1, I_2, I_3 are the principal invariants of the strain tensor C_{AB}, i.e.

$$I_1 = \text{tr}\,\underline{C},\ I_2 = \frac{1}{2}[(\text{tr}\,\underline{C})^2 - \text{tr}\,\underline{C}^2],\ I_3 = J^2, \qquad (1.3.2)$$

where $\underline{C} = (C_{AB})$ and $\text{tr}\,\underline{C} = C_{AA}$.

It follows from (1.1.11), (1.1.12) and (1.3.2) that

$$\frac{\partial I_1}{\partial E_{AB}} = 2\delta_{AB},\quad \frac{\partial I_2}{\partial E_{AB}} = 2\Lambda_{AB},\quad \frac{\partial I_3}{\partial E_{AB}} = 2I_3C^{(-1)}_{AB}, \qquad (1.3.3)$$

where

$$\Lambda_{AB} = I_1 \ \delta_{AB} - C_{AB}, \quad C_{AB}^{(-1)} = X_{A,i}X_{B,j}. \tag{1.3.4}$$

Thus, we obtain

$$T_{AB} = \gamma_{-1}C_{AB}^{(-1)} + \gamma_o \ \delta_{AB} + \gamma_1 C_{AB}, \tag{1.3.5}$$

where

$$\gamma_{-1} = 2I_3 \frac{\partial e}{\partial I_3}, \quad \gamma_o = 2(\frac{\partial e}{\partial I_1} + I_1 \frac{\partial e}{\partial I_2}), \ \gamma_1 = -2 \frac{\partial e}{\partial I_2}. \tag{1.3.6}$$

Next, let

$$\Phi = \frac{2}{\sqrt{I_3}} \frac{\partial e}{\partial I_1}, \ \Psi = \frac{2}{\sqrt{I_3}} \frac{\partial e}{\partial I_2}, \quad p = 2\sqrt{I_3} \frac{\partial e}{\partial I_3}. \tag{1.3.7}$$

Then (1.3.6) implies the alternative form of (1.3.5)

$$T_{AB} = \sqrt{I_3} \ \ (\Phi\delta_{AB} + \Psi\Lambda_{AB} + p \ C_{AB}^{(-1)}). \tag{1.3.8}$$

Clearly,

$$\gamma_{-1} = \sqrt{I_3} \ p, \ \gamma_o = \sqrt{I_3} \ \Phi + I_1 \sqrt{I_3} \ \Psi, \ \gamma_1 = -\sqrt{I_3} \ \Psi. \tag{1.3.9}$$

It follows from (1.1.12), (1.1.19) and (1.3.5) that

$$t_{ij} = \alpha_o \ \delta_{ij} + \alpha_1 b_{ij} + \alpha_2 b_{is} b_{sj}, \tag{1.3.10}$$

where

$$\alpha_{h+1} = \frac{1}{\sqrt{I_3}} \gamma_h \quad (h = -1,0,1).$$

On the other hand, if we denote

$$B_{ij} = x_{i,A} x_{j,B} \Lambda_{AB} = I_1 b_{ij} - b_{is} b_{sj}, \tag{1.3.11}$$

then (1.1.19) and (1.3.8) imply that

$$t_{ij} = \Phi b_{ij} + \Psi B_{ij} + p\delta_{ij}. \tag{1.3.12}$$

Our next objective is to determine the form of the coefficients A_{MNKL} and C_{ijrs} which appear in the relations (1.2.25) and (1.2.33). By (1.1.15), (1.2.20) and (1.3.5) we have

$$A_{ABMN} = \frac{1}{2}\left(\frac{\partial}{\partial E_{MN}} + \frac{\partial}{\partial E_{NM}}\right)\left(\gamma_{-1} C^{(-1)}_{AB} + \gamma_o\ \delta_{AB} + \gamma_1 C_{AB}\right). \tag{1.3.13}$$

By using the relations (1.3.3) and

$$\frac{\partial C^{(-1)}_{AB}}{\partial E_{MN}} = -2\ C^{(-1)}_{AM}\ C^{(-1)}_{BN}, \tag{1.3.14}$$

it follows form (1.3.13) that

$$A_{ABMN} = A^{(-1)}_{MN} C^{(-1)}_{AB} + A^{(o)}_{MN}\ \delta_{AB} + A^{(1)}_{MN} C_{AB} -$$

$$-\gamma_{-1}\left(C^{(-1)}_{AM} C^{(-1)}_{BN} + C^{(-1)}_{AN} C^{(-1)}_{BM}\right) + \gamma_1\left(\delta_{AM}\ \delta_{BN} + \delta_{AN}\ \delta_{BM}\right), \tag{1.3.15}$$

where

$$A^{(h)}_{MN} = 2\left(\frac{\partial\gamma_h}{\partial I_1}\ \delta_{MN} + \frac{\partial\gamma_h}{\partial I_2}\ \Lambda_{MN} + \frac{\partial\gamma_h}{\partial I_3}\ I_3 C^{(-1)}_{MN}\right)\ (h = -1,0,1). \tag{1.3.16}$$

In view of the relations (1.2.30) and (1.3.15) we are led to the relation

$$C_{ijrs} = \delta_{ij}\ \alpha^{(-1)}_{rs} + b_{ij}\ \alpha^{(o)}_{rs} + b_{ip}b_{jp}\ \alpha^{(1)}_{rs} - \tag{1.3.17}$$

$$- p(\delta_{ir}\ \delta_{js} + \delta_{is}\ \delta_{jr}) - \Psi(b_{ir}b_{js} + b_{is}b_{jr}),$$

where

$$\alpha^{(i)}_{rs} = \frac{1}{J}\ x_{r,M}x_{s,N}A^{(i)}_{MN} = \frac{2}{J}\ \left(\frac{\partial\gamma_i}{\partial I_1}\ b_{rs} + \frac{\partial\gamma_i}{\partial I_2}\ B_{rs} + \right.$$

$$\left. + I_3\ \frac{\partial\gamma_i}{\partial I_3}\ \delta_{rs}\right), \qquad (i = -1,0,1). \tag{1.3.18}$$

If we now define

$$A_{ij} = \frac{2}{J}\frac{\partial^2 e}{\partial I_i \partial I_j}, \tag{1.3.19}$$

then the relation (1.3.18) implies

$$\alpha_{rs}^{(-1)} = 2I_3[A_{13}b_{rs} + A_{23}B_{rs} + (\frac{p}{I_3} + A_{33}I_3)\delta_{rs}],$$

$$\alpha_{rs}^{(o)} = 2[(A_{11} + I_1A_{12} + \Psi)b_{rs} + (A_{12} + I_1A_{22})B_{rs} +$$

$$+ (A_{13} + I_1A_{23})I_3\,\delta_{rs}], \tag{1.3.20}$$

$$\alpha_{rs}^{(1)} = -\,2(A_{12}b_{rs} + A_{22}B_{rs} + A_{23}I_3\,\delta_{rs}).$$

If the continuum is incompressible then $I_3 = 1$ and $\rho = \rho_0$. Introducing the scalar function p as a Lagrange multiplier to incorporate this constraint, with the aid of (1.3.3) we obtain, in place of (1.1.15),

$$T_{AB} = p\,C_{AB}^{(-1)} + \frac{\partial e}{\partial E_{AB}}.$$

The analysis in terms of infinitesimal superimposed deformations is entirely similar to the one in compressible materials. We have now $J = J^* = 1$ so that the incremental displacement fields are subject to the incompressibility condition

$$u_{i,i} = 0.$$

By using the procedure given in Section 1.2 we obtain

$$s_{ji} = p'\,\delta_{ij} - p(u_{i,j} + u_{j,i}) + d^o_{ijrs}u_{r,s}, \tag{1.3.21}$$

where p' is an arbitrary scalar function of position and time and

$$d^o_{ijrs} = C^o_{ijrs} + \delta_{ir}t_{js},\quad C^o_{ijrs} = x_{i,K}x_{j,L}x_{r,M}x_{s,N}A_{KLMN}. \tag{1.3.22}$$

If the material is incompressible and isotropic, we only know the value of the internal energy function in terms of I_1 and I_2 when $I_3 = 1$. Thus, $e = \hat{e}(I_1, I_2)$. The stress-strain equations (1.3.12) retain the same form

$$t_{ij} = p\,\delta_{ij} + \Phi^o b_{ij} + \Psi^o B_{ij}; \tag{1.3.23}$$

where

$$\Phi^o = 2\frac{\partial \hat{e}}{\partial I_1}, \quad \Psi^o = 2\frac{\partial \hat{e}}{\partial I_2}, \tag{1.3.24}$$

and p is the Lagrange multiplier. The coefficients C^o_{ijrs} become

$$C^o_{ijrs} = -\Psi^o(b_{ir}b_{js} + b_{is}b_{jr}) + b_{ij}\,\hat{\alpha}^{(o)}_{rs} + b_{ip}b_{jp}\hat{\alpha}^{(1)}_{rs}, \tag{1.3.25}$$

where

$$\hat{\alpha}^{(g)}_{rs} = 2\left(\frac{\partial \gamma_g}{\partial I_1} b_{rs} + \frac{\partial \gamma_g}{\partial I_2} B_{rs}\right), \quad (g = 0,1), \tag{1.3.26}$$

$$\hat{\gamma}_o = \Phi^o + I_1\Psi^o, \quad \gamma_1 = -\Psi^o.$$

If we now define

$$A^o_{\alpha\beta} = 2\frac{\partial^2 \hat{e}}{\partial I_\alpha \partial I_\beta}, \tag{1.3.27}$$

then

$$\begin{aligned} \hat{\alpha}^{(o)}_{rs} &= 2[(A^o_{11} + I_1A^o_{12} + \Psi^o)b_{rs} + (A^o_{12} + I_1A^o_{22})B_{rs}], \\ \hat{\alpha}^{(1)}_{rs} &= -2(A^o_{12}b_{rs} + A^o_{22}B_{rs}). \end{aligned} \tag{1.3.28}$$

1.4 The classical theory of elastic bodies with initial stress

Following Green [137], we now assume that the primary state B is identical with that of the initial body B_0 so that $x_1 = X_1$, $x_2 = X_2$, and $x_3 = X_3$. By an initial stress field we mean a stress field on B_0 which satisfies the equilibrium equations. We may not know how the body acquired this stress.

We suppose that the body B_o is subjected to initial stress. Then, we consider the configuration B* which is obtained from B (or B_o) by a small deformation. We have

$$J = 1, \ \rho_o = \rho, \ E_{AB} = 0, \ T_{AB} = \frac{\partial e}{\partial E_{AB}}, \tag{1.4.1}$$

the derivatives being evaluated at $E_{AB} = 0$. Clearly, the stress tensors T_{Ai}, T_{AB} and t_{ij} coincide with the initial stress. The coefficients C_{ijrs} are now evaluated at $E_{AB} = 0$.

The work of Section 1.2 can be applied to this special case and yields the linear theory of elastic bodies which are under initial stress. The equations of this theory are the equations of motion

$$s_{ji,j} + \rho F_i = \rho \ddot{u}_i \quad \text{on} \quad B_o \times (t_o, t_1), \tag{1.4.2}$$

and the "constitutive" equations

$$s_{ji} = (C_{ijrs} + t_{js}\delta_{ir})u_{r,s}, \tag{1.4.3}$$

where ρ is now the density of the initial body and t_{ij} is the initial stress. If the initial body B_o is homogeneous then C_{ijrs} are constants.

To the equations (1.4.2) and (1.4.3) we adjoin the boundary conditions (1.2.46) and the initial conditions (1.2.41).

The value of the energy e* which corresponds to the theory of this section may be written as a special case of (1.2.50) when $E_{AB} = 0$, and is

$$\begin{aligned} e^* &= e_o + t_{ij}u_{i,j} + \frac{1}{2}(C_{ijrs}u_{i,j}u_{r,s} + t_{rs}u_{i,r}u_{i,s}) = \\ &= e_o + t_{ij}e_{ij} + \frac{1}{2}(C_{ijrs}e_{ij}e_{rs} + t_{rs}u_{i,r}u_{i,s}), \end{aligned} \tag{1.4.4}$$

where e_o is constant if B_o is homogeneous. This expression is a function not only of classical strain e_{ij}, but also of all displacement gradients $u_{r,s}$ in view of the presence of initial stress t_{ij}. If we put

$$u_{i,j} = e_{ij} + \omega_{ij}, \ \omega_{ij} = \frac{1}{2}(u_{i,j} - u_{j,i}), \tag{1.4.5}$$

then the energy e^* may be expressed as a function of classical strains e_{ij} and rotations ω_{ij}.

If the body B_o is isotropic then

$$t_{ij} = p\,\delta_{ij},\ C_{ijrs} = \lambda\delta_{ij}\,\delta_{rs} + \mu(\delta_{ir}\,\delta_{js} + \delta_{is}\,\delta_{jr}), \qquad (1.4.6)$$

where p, λ, μ are specified functions of x_i. It follows that the only initial stress possible in an isotropic body is hydrostatic pressure or tension. If the body B_o is homogeneous then the coefficients p, λ and μ are constants. The constants λ and μ are not necessarily the classical Lamé constants for an isotropic solid with zero initial stress. Clearly, an initially stressed isotropic body can be regarded as a material requiring three coefficients for its specification: λ, μ and p.

1.5 Infinitesimal initial deformations

In this section we consider the special case of the general theory of Section 1.2 when the primary state B is obtained from B_o by an infinitesimal deformation (cf. Green [137]). We assume that the body B_o is unstressed.

Since the coefficients A_{KLMN} given by (1.2.20) are found from second derivatives of the internal energy e corresponding to the primary state, it is sufficient to consider a cubic for e of the form

$$e = e_o + \frac{1}{2} c_{ABCD}E_{AB}E_{CD} + \frac{1}{6} c_{ABCDMN}E_{AB}E_{CD}E_{MN}, \qquad (1.5.1)$$

where e_o, c_{ABCD}, c_{ABCDMN} are, in general, functions of the coordinates X_A. Linear terms in E_{AB} are omitted since B_o is unstressed. The coefficients c_{ABCD} and c_{ABCDMN} satisfy the symmetry relations

$$\begin{aligned} c_{ABCD} &= c_{CDAB} = c_{BACD},\ c_{ABCDMN} = c_{BACDMN} = c_{ABDCMN} = \\ &= c_{ABCDNM} = c_{CDABMN} = c_{ABMNCD} = c_{MNCDAB}. \end{aligned} \qquad (1.5.2)$$

If the body B_o is homogeneous then the coefficients e_o, c_{ABCD}, c_{ABCDMN} are constants.

We now let

$$v_i = x_i - \delta_{iA}X_A, \tag{1.5.3}$$

where δ_{iA} is the Kronecker delta. We assume that $v_i = \varepsilon'' v_i'$ where ε'' is a constant which is small so that squares and higher powers of ε'' may be neglected compared with ε'', and v_i' is independent of ε''.

It follows from (1.1.15), (1.2.20) and (1.5.1) that

$$T_{AB} = c_{ABMN}\varepsilon_{MN}, \tag{1.5.4}$$

$$A_{KLMN} = c_{KLMN} + c_{KLMNRS}\varepsilon_{RS},$$

where

$$\varepsilon_{AB} = \frac{1}{2}(v_{A,B} + v_{B,A}), \quad v_A = \delta_{iA}v_i, \tag{1.5.5}$$

and only linear terms in the displacement gradients are retained in (1.5.4).

With the aid of (1.2.25), (1.5.3) and (1.5.4) we obtain

$$E_{iAjN} = c_{ABMN}(\delta_{iB}\delta_{jM} + \delta_{jM}v_{i,B} + \delta_{iB}v_{j,M}) + \tag{1.5.6}$$

$$+ \delta_{iB}\,\delta_{jM}c_{ABMNRS}\varepsilon_{RS}.$$

From (1.1.20) and (1.5.3),

$$J = 1 + v_{A,A}, \tag{1.5.7}$$

if we retain only linear terms in the displacement gradients. Also, from (1.2.30), (1.5.3), (1.5.4) and (1.5.7) we can determine the coefficients C_{ijrs}.

We now suppose that the continuum in the state B_o is isotropic. Then, we have

$$c_{KLMN} = \lambda\delta_{KL}\delta_{MN} + \mu(\delta_{KM}\delta_{LN} + \delta_{KN}\delta_{LM}),$$

$$\begin{aligned} c_{KLMNRS} = {} & \nu_1 \delta_{KL}\delta_{MN}\delta_{RS} + \nu_2[\delta_{KL}(\delta_{MR}\delta_{NS} + \delta_{MS}\delta_{NR}) + \\ & + \delta_{MN}(\delta_{KR}\delta_{LS} + \delta_{KS}\delta_{LR}) + \delta_{RS}(\delta_{KM}\delta_{LN} + \\ & + \delta_{KN}\delta_{LM})] + \nu_3[\delta_{KM}(\delta_{LR}\delta_{NS} + \delta_{LS}\delta_{NR}) + \\ & + \delta_{LN}(\delta_{KR}\delta_{MS} + \delta_{KS}\delta_{MR}) + \delta_{KN}(\delta_{LR}\delta_{MS} + \\ & + \delta_{LS}\delta_{MR}) + \delta_{LM}(\delta_{KR}\delta_{NS} + \delta_{KS}\delta_{NR})]. \end{aligned} \tag{1.5.8}$$

The constitutive coefficients λ, μ, and ν_i are constants if the body B_0 is homogeneous. The coefficients ν_i have been introduced by Toupin and Bernstein [350]. It follows from (1.5.6) and (1.5.8) that

$$\begin{aligned} E_{iAjN} = {} & (\lambda + \nu_1\gamma)\delta_{iA}\delta_{jN} + (\mu + \nu_2\gamma)(\delta_{jA}\delta_{iN} + \\ & + \delta_{ij}\delta_{AN}) + (\lambda v_{M,N} + 2\nu_2\varepsilon_{MN})\delta_{iA}\delta_{jM} + \\ & + (\lambda v_{L,A} + 2\nu_2\varepsilon_{AL})\delta_{iL}\delta_{jN} + (\mu v_{L,N} + \\ & + 2\nu_3\varepsilon_{LN})\delta_{jA}\delta_{iL} + (\mu v_{M,A} + 2\nu_3\varepsilon_{AM})\delta_{iN}\delta_{jM} + \\ & + 2(\mu + \nu_3)\delta_{AN}\varepsilon_{PQ}\delta_{iP}\delta_{jQ}, \end{aligned} \tag{1.5.9}$$

where $\gamma = \varepsilon_{RR}$. Also, from (1.2.30), (1.5.3), (1.5.4), (1.5.7) and (1.5.8), we obtain

$$\begin{aligned} C_{ijrs} = {} & [\lambda + (\nu_1 - \lambda)\gamma]\delta_{ij}\delta_{rs} + \\ & + [\mu + (\nu_2 - \mu)\gamma](\delta_{ir}\delta_{js} + \delta_{is}\delta_{jr}) + \\ & + 2(\lambda + \nu_2)(\delta_{ij}\varepsilon_{rs} + \delta_{rs}\varepsilon_{ij}) + \\ & + 2(\mu+\nu_3)(\delta_{ir}\varepsilon_{js} + \delta_{js}\varepsilon_{ir} + \delta_{is}\varepsilon_{jr} + \\ & + \delta_{jr}\varepsilon_{is}), \end{aligned} \tag{1.5.10}$$

where $\varepsilon_{ij} = \delta_{iA}\,\delta_{jB}\,\varepsilon_{AB}$.

If the body B_0 has rhombic symmetry with respect to the rectangular Cartesian axes X_A, then e is restricted to the functional form (see, for example, Green and Zerna [140], Sect. 5.4)

$$e = e(E_{11},E_{22},E_{33},E_{12}^2,E_{24}^2,E_{31}^2,E_{12}E_{23}E_{31}). \tag{1.5.11}$$

It follows from (1.5.1) and (1.5.11) that only non-zero coefficients c_{ABCD} are

$$\begin{matrix} c_{1111}, & c_{2222}, & c_{3333}, & c_{1122}, & c_{1133}, \\ c_{2233}, & c_{1313}, & c_{2323}, & c_{1212}, & \end{matrix} \tag{1.5.12}$$

if we remember the symmetry conditions (1.5.2). The only non-zero coefficients c_{ABCDMN} are

$$c_{(KLMN)11},\ c_{(KLMN)22},\ c_{(KLMN)33},\ c_{122331}, \tag{1.5.13}$$

where $(KLMN) \in \{(1111), (2222), (3333), (1122), (1133), (2233), (1313), (2323), (1212)\}$.

2 Elastostatics

2.1 Boundary-value problems

We assume that the body is in equilibrium in both states B and B*. The fundamental system of field equations for the time-independent behaviour of an elastic body, referred to the primary state, consists of the stress-displacement relations

$$s_{ji} = d_{ijrs}u_{r,s}, \tag{2.1.1}$$

and the equations of equilibrium

$$s_{ji,j} + \rho F_i = 0, \tag{2.1.2}$$

on B. The coefficients d_{ijrs} are functions of the state of strain of the body in its primary state. Since B is in equilibrium, d_{ijrs} and ρ are time-independent. We assume that

(i) d_{ijrs} are continuously differentiable on $\bar{B}$ and satisfy the symmetry relations (1.2.45);

(ii) ρ is continuous and strictly positive on $\bar{B}$;

(iii) $\underline{F}$ is continuous on $\bar{B}$.

We say that the vector field $\underline{u}$ is an admissible displacement field on $\bar{B}$ provided $\underline{u} \in C^2(B) \cap C^1(\bar{B})$. An admissible stress field on $\bar{B}$ is a tensor field s_{ij} with the following properties: (i) s_{ij} is smooth on B; (ii) s_{ij} and $s_{ij,i}$ are continuous on $\bar{B}$. By an admissible state on $\bar{B}$ we mean an ordered array of functions $A = (u_i, s_{mn})$ with the properties: (i) $\underline{u}$ is an admissible displacement field on $\bar{B}$; (ii) s_{ij} is an admissible stress field on $\bar{B}$. We say that $A = (u_i, s_{mn})$ is an elastic state on $\bar{B}$ corresponding to the body force field $\underline{F}$ if A is an admissible state that satisfies the stress-displacement relations (2.1.1) and the equations of equilibrium (2.1.2). The corresponding surface traction $\underline{s}$ is then defined at every regular point of ∂B by

$$s_i = s_{ji}n_j. \tag{2.1.3}$$

When we omit mention of the domain of definition of an elastic state, it will always be understood to be $\bar{B}$.

The mixed problem of elastostatics is to find an elastic state (u_i, s_{pq}) that corresponds to the body force $\underline{F}$ and satisfies the boundary conditions

$$\underline{u} = \underline{\tilde{u}} \text{ on } \bar{S}_1, \quad \underline{s} = \underline{\tilde{p}} \text{ on } S_2, \tag{2.1.4}$$

where $\underline{\tilde{u}}$ and $\underline{\tilde{p}}$ are given vector fields. If S_2 is empty (i.e. $S_1 = \partial B$), the boundary conditions (2.1.4) reduce to

$$\underline{u} = \underline{\tilde{u}} \text{ on } \partial B, \tag{2.1.5}$$

and the resulting problem is called the displacement problem. When $S_2 = \partial B$, the boundary conditions become

$$\underline{s} = \underline{\tilde{p}} \text{ on } \partial B, \tag{2.1.6}$$

and the associated problem is called the traction problem. In this case, for equilibrium, the loading is constrained by

$$\int_B \rho F_i dv + \int_{\partial B} \tilde{p}_i da = 0. \tag{2.1.7}$$

We assume that: (α) $\underline{\tilde{u}}$ is continuous on $\bar{S}_1$; (β) $\underline{\tilde{p}}$ is piecewise regular on S_2; (γ) $\underline{F}$ is continuous on $\bar{B}$. By an external data system on $\bar{B}$ for the mixed problem we mean an ordered array $L = (\underline{F}, \underline{\tilde{u}}, \underline{\tilde{p}})$ with the properties (α) - (γ).

Substitution of (2.1.1) into the equations (2.1.2) and (2.1.4) yields the equations

$$(d_{ijrs}u_{r,s})_{,j} + \rho F_i = 0 \quad \text{on} \quad B, \tag{2.1.8}$$

and the boundary conditions

$$u_i = \tilde{u}_i \text{ on } \bar{S}_1, \quad d_{ijrs}u_{r,s}n_j = \tilde{p}_i \text{ on } S_2. \tag{2.1.9}$$

By a solution of the mixed problem of elastostatics we mean an admissible displacement field $\underline{u}$ on $\bar{B}$ that satisfies the equations (2.1.8) and the boundary conditions (2.1.9).

We now assume that the deformation of the body is referred to the configuration B_o. In this case the fundamental system of field equations for the time-independent behaviour of an elastic body consists of the stress-displacement relations

$$S_{Ai} = D_{iAjK}u_{j,K}, \tag{2.1.10}$$

and the equations of equilibrium

$$S_{Ai,A} + \rho_o F_i = 0, \tag{2.1.11}$$

on B_o. We assume that D_{iAjK} are continuously differentiable on B_o and satisfy the symmetry relations (1.2.38), and that the density ρ_o is continuous and strictly positive on $\bar{B}_o$. Let (u_i, S_{Aj}) be an elastic state on $\bar{B}_o$. The associated surface traction $\underline{P}$ is then defined at every regular point of ∂B_o by

$$P_i = S_{Ai}N_A. \tag{2.1.12}$$

The mixed problem of elastostatics consists in finding an elastic state (u_i, S_{Aj}) on $\bar{B}_o$ that corresponds to the body force $\underline{F}$ and satisfies the boundary conditions

$$\underline{u} = \hat{\underline{u}} \text{ on } \bar{\Sigma}_1, \quad \underline{P} = \hat{\underline{P}} \text{ on } \Sigma_2, \tag{2.1.13}$$

where $\hat{\underline{u}}$ and $\hat{\underline{P}}$ are specified vector fields. Substitution of (2.1.10) into (2.1.11) and (2.1.13) yields the equations

$$(D_{iAjK}u_{j,K})_{,A} + \rho_o F_i = 0 \quad \text{on } B_o, \tag{2.1.14}$$

and the boundary conditions

$$u_i = \hat{u}_i \text{ on } \bar{\Sigma}_1, \quad D_{iAjK}u_{j,K}N_A = \hat{P}_i \text{ on } \Sigma_2. \tag{2.1.15}$$

By a solution of the mixed problem of elastostatics we now mean an admissible displacement field $\underline{u}$ on $\bar{B}_o$ that satisfies the equations (2.1.14) and the boundary condition (2.1.15).

When Σ_1 is empty the loading is constrained by

$$\int_{B_o} \rho_o F_i dV + \int_{\partial B_o} \hat{P}_i dA = 0. \tag{2.1.16}$$

Remark. Let us consider the equilibrium of moments. For the resultant moment of the surface tractions and body forces in B*, we find

$$e_{ijk}[\int_{\partial B} (x_j+u_j)(t_{rk}+s_{rk})n_r da + \int_B \rho(x_j+u_j)(f_k+F_k)dv] = 0, \tag{2.1.17}$$

or

$$e_{ijk}[\int_{\partial B_o} (x_j+u_j)(T_{Lk}+S_{Lk})N_L dA + \int_{B_o} \rho_o(x_j+u_j)(f_k+F_k)dV] = 0, \tag{2.1.18}$$

where e_{ijk} is the permutation symbol. Since the primary state is also in equilibrium we have

$$e_{ijk}[\int_{\partial B} x_j t_{rk} n_r da + \int_B \rho\, x_j f_k dv] = 0, \tag{2.1.19}$$

or

$$e_{ijk}[\int_{\partial B_o} x_j T_{Lk} N_L dA + \int_{B_o} \rho_o x_j f_k dV] = 0. \tag{2.1.20}$$

By (2.1.2), (2.1.17) and (2.1.19), we conclude that

$$e_{ijk}[\int_{\partial B} (u_j t_{rk}+x_j s_{rk})n_r da + \int_B \rho(u_j f_k+x_j F_k)dv] = 0. \tag{2.1.21}$$

On the other hand, from (2.1.11), (2.1.18) and (2.1.20) we get

$$e_{ijk}[\int_{\partial B_o} (u_j T_{Lk}+x_j S_{Lk})N_L dA + \int_{B_o} \rho_o(u_j f_k+x_j F_k)dV] = 0. \tag{2.1.22}$$

If (u_i, s_{mn}) is an elastic state on $\bar{B}$ corresponding to the body force $\underline{F}$, then by the divergence theorem we can rewrite (2.1.21) as follows

$$e_{ijk} \int_B (u_{j,r} t_{rk} + s_{jk}) dv = 0. \qquad (2.1.23)$$

In view of (1.2.31), (1.2.43) and (1.2.44) we conclude that (2.1.23) is satisfied. Clearly, when $S_2 = \partial B$, the relations (1.1.18), (2.1.6) and (2.1.21) imply the moment conditions

$$e_{ijk}[\int_{\partial B} (u_j \tilde{t}_k + x_j \tilde{p}_k) da + \int_B \rho(u_j f_k + x_j F_k) dv] = 0. \qquad (2.1.24)$$

The relations (2.1.24) are to be regarded as conditions of compatibility to be satisfied by any solution of the traction boundary-value problem. An alternative form of the moment conditions can be derived using (2.1.22). The moment conditions have been employed by Beatty [21], [22] in a discussion of elastic stability (see also Truesdell and Noll [356], Sect. 68 b). We now record, without proof, the following well-known result (Beatty [21], Truesdell and Noll [356], Sect. 68 b): if the loads acting on the primary state possess no axis of equilibrium, the moment conditions determine the infinitesimal rotations in the traction boundary-value problem; if there is an axis of equilibrium, only rotations about that axis are arbitrary.

2.2 The reciprocal theorem. Variational theorems

In this section we shall establish the reciprocal theorem and discuss some variational theorems.

When we discuss general theorems of elastostatics we will frequently take the primary state as reference configuration. Since the elasticities depend upon the initial strain, the theory of infinitesimal deformations superimposed on finite deformations has given impetus to researches of the equations of elastic bodies for which little information is known concerning elasticities.

The derivation of the reciprocal theorem in the equilibrium theory of initially strained bodies is contained in the works of Zorski [389] and Zubov [390] (see also Truesdell and Noll [356], Sect. 88). In essence, the reciprocal theorem expresses the fact that the underlying system of field equations is self-adjoint.

Theorem 2.2.1. Let $(u_i^{(1)}, s_{mn}^{(1)})$ and $(u_i^{(2)}, s_{mn}^{(2)})$ be elastic states corresponding to the external data systems $(F_i^{(1)}, \tilde{u}_i^{(1)}, \tilde{p}_i^{(1)})$ and $(F_i^{(2)}, \tilde{u}_i^{(2)}, \tilde{p}_i^{(2)})$, respectively. Then

$$\int_B \rho F_i^{(1)} u_i^{(2)} dv + \int_{\partial B} s_i^{(1)} u_i^{(2)} da = \int_B \rho F_i^{(2)} u_i^{(1)} dv + \int_{\partial B} s_i^{(2)} u_i^{(1)} da, \tag{2.2.1}$$

where

$$s_i^{(\alpha)} = s_{jk}^{(\alpha)} n_j \quad (\alpha = 1,2) \text{ on } \partial B.$$

Proof. By (1.2.45) and (2.1.1),

$$s_{ji}^{(1)} u_{i,j}^{(2)} = s_{ji}^{(2)} u_{i,j}^{(1)} . \tag{2.2.2}$$

Let

$$K_{\alpha\beta} = \int_B s_{ji}^{(\alpha)} u_{i,j}^{(\beta)} dv. \tag{2.2.3}$$

It follows from (2.2.2) and (2.2.3) that

$$K_{12} = K_{21}. \tag{2.2.4}$$

By the divergence theorem and (2.1.2),

$$K_{\alpha\beta} = \int_{\partial B} s_i^{(\alpha)} u_i^{(\beta)} da + \int_B \rho F_i^{(\alpha)} u_i^{(\beta)} dv. \tag{2.2.5}$$

From (2.2.4) and (2.2.5) we obtain the desired result. □

By (2.2.3) and (2.2.5), it follows that

$$\int_B s_{ji} u_{i,j} dv = \int_B \rho F_i u_i dv + \int_{\partial B} s_i u_i da.$$

Thus, we have

$$\int_B d_{ijrs} u_{r,s} u_{i,j} dv = \int_B \rho F_i u_i dv + \int_{\partial B} s_i u_i da, \tag{2.2.6}$$

for every admissible displacement field $\underline{u}$ on $\bar{B}$ that satisfies the equations (2.1.8). The above relation extends the theorem of work and energy from the classical elastostatics (see, for example, Gurtin [153], p. 95).

Let M denote the set of all admissible displacement fields on $\bar{B}$ which satisfy the boundary conditions

$$u_i = 0 \text{ on } \bar{S}_1, \quad d_{ijrs}u_{r,s}n_j = 0 \text{ on } S_2. \tag{2.2.7}$$

Let $\underline{A}$ be the operator on M defined by

$$A_i\underline{u} = -(d_{ijrs}u_{r,s})_{,j}. \tag{2.2.8}$$

Clearly, the field equations (2.1.8) can be written in the form

$$\underline{Au} = \underline{f} \text{ on } B, \tag{2.2.9}$$

where $\underline{f} = (\rho F_i)$. Let $\underline{u},\underline{v} \in M$. Then (2.2.1) implies

$$\int_B (\underline{u}\cdot\underline{Av} - \underline{v}\cdot\underline{Au})dv = 0, \tag{2.2.10}$$

so that $\underline{A}$ is symmetric. When the tensor field d_{ijrs} is positive definite, then it is a simple matter to derive existence theorems following the work of Fichera [120] (see also Zorski [389]). Moreover, similarly to the classical elasticity it is possible to apply variational methods to all basic boundary-value problems (see, for example, Mikhlin [260]).

Variational theorems for the theory of small deformations superimposed on finite deformation are given by Zorski [389], Shield and Fosdick [323] and Zubov [390].

With a view toward deriving a variational theorem, we now state, without proof the following result (Gurtin [153], Sect. 35).

Lemma 2.2.1. Let W be a finite-dimensional inner product space. Let $\underline{w} : S_2 \to W$ be piecewise regular and satisfy

$$\int_{S_2} \underline{w}\cdot\underline{v}\, da = 0,$$

for every class C^∞ function $\underline{v} : \bar{B} \to W$ that vanishes near S_1. Then $\underline{w} = \underline{0}$ on S_2.

Let L be a linear space, K a subset of L, and $F(\cdot)$ a functional defined on K. Let

$$y, y^0 \in L, \; y + \lambda y^0 \in K \text{ for every scalar } \lambda, \tag{2.2.11}$$

and define

$$\delta_{y^0} F(y) = \frac{d}{d\lambda} F(y + \lambda y^0)\Big|_{\lambda = 0}.$$

We say that the variation of $F(\cdot)$ is zero at $y \in K$ and write

$$\delta F(y) = 0,$$

if $\delta_{y^0} F(y)$ exists and equals zero for every choice of y^0 consistent with (2.2.11).

By a kinematically admissible displacement field we mean an admissible displacement field on $\bar{B}$, that satisfies the displacement boundary condition.

Theorem 2.2.2. Let V denote the set of all kinematically admissible displacement fields, and let G be the functional on V defined by

$$G(\underline{u}) = \int_B \left(\frac{1}{2} d_{ijrs} u_{r,s} u_{i,j} - \rho F_i u_i\right) dv - \int_{S_2} \tilde{p}_i u_i \, da,$$

for every $\underline{u} \in V$. Then

$$\delta G(\underline{u}) = 0 \tag{2.2.12}$$

at $\underline{u} \in V$ if and only if $\underline{u}$ is a solution of the mixed problem.

Proof. Let $\underline{u}^0$ be an admissible displacement field on $\bar{B}$, and assume that $\underline{u} + \lambda \underline{u}^0 \in V$ for every scalar λ. This latter condition is equivalent to the requirement that $\underline{u}^0$ meets the boundary condition

$$\underline{u}^o = \underline{0} \text{ on } \bar{S}_1. \tag{2.2.13}$$

In view of the symmetry of d_{ijrs},

$$\delta_{\underline{u}^o} G(\underline{u}) = \int_B (d_{ijrs} u_{r,s} u^o_{i,j} - \rho F_i u^o_i) dv - \int_{S_2} \tilde{p}_i u^o_i da.$$

Next, using the divergence theorem and (2.2.13), we obtain

$$\begin{aligned} \delta_{\underline{u}^o} G(\underline{u}) = &- \int_B [(d_{ijrs} u_{r,s})_{,j} + \rho F_i] u^o_i dv + \\ &+ \int_{S_2} (d_{ijrs} u_{r,s} n_j - \tilde{p}_i) u^o_i da, \end{aligned} \tag{2.2.14}$$

for every admissible displacement field $\underline{u}^o$ that satisfies (2.2.13). If $\underline{u}$ is a solution to the mixed problem, then (2.2.14) yields

$$\delta_{\underline{u}^o} G(\underline{u}) = 0, \tag{2.2.15}$$

for every admissible displacement field $\underline{u}^o$ that satisfies (2.2.13). Thus we conclude that (2.2.12) holds. Conversely, suppose that (2.2.12) and hence (2.2.15) holds. Let $\underline{u}^o$ vanish near ∂B. Then it follows from (2.2.14) and (2.2.15) that

$$\int_B [(d_{ijrs} u_{r,s})_{,j} + \rho F_i] u^o_i dv = 0.$$

Since this relation must hold for every such $\underline{u}^o$ of class C^1 on $\bar{B}$, we conclude from the fundamental lemma that (2.1.8) holds. By (2.2.14), (2.2.15) and (2.1.8),

$$\int_{S_2} (d_{ijrs} u_{r,s} n_j - \tilde{p}_i) u^o_i da = 0,$$

for every admissible displacement field $\underline{u}^o$ that satisfies (2.2.13). Then, Lemma 2.2.1 implies $d_{ijrs} u_{r,s} n_j = \tilde{p}_i$ on S_2. Thus $\underline{u}$ is a solution of the mixed problem. □

When the tensor field d_{ijrs} is positive definite we are led to the following result.

Theorem 2.2.3. Assume that the tensor field d_{ijrs} is positive definite. Let $\underline{u}$ be a solution of the mixed problem. Then $G(\underline{u}) \leqq G(\underline{u}^*)$, for every kinematically admissible displacement field $\underline{u}^*$.

This theorem is a counterpart of the theorem of minimum potential energy in classical elastostatics (see Gurtin [153], Sect. 34).

2.3 Uniqueness theorems

In this section we discuss the uniqueness question appropriate to the basic boundary-value problems of elastostatics. We assume that the body occupies a bounded regular region of three-dimensional Euclidean space. The following result is a direct counterpart of Kirchhoff's uniqueness theorem in the classical elastostatics (cf. Knops and Payne [227]).

Theorem 2.3.1. Let the tensor field d_{ijrs} be positive definite in the sense that there exists a positive constant d such that

$$\int_B d_{ijrs} v_{ij} v_{rs} dv \geqq d \int_B v_{ij} v_{ij} dv, \tag{2.3.1}$$

for all tensors v_{ij}. Then

(i) any two solutions of the traction boundary-value problem are equal modulo a rigid body translation;

(ii) the displacement boundary-value problem has at most one solution.

Proof. Suppose that there are two solutions $\underline{u}^{(1)}$ and $\underline{u}^{(2)}$ of the same problem. Then, the difference $\underline{w} = \underline{u}^{(1)} - \underline{u}^{(2)}$ corresponds to zero body force and satisfies

$$\underline{w} \cdot \underline{s}(\underline{w}) = 0 \text{ on } \partial B.$$

Thus, by (2.2.6) we obtain

$$\int_B d_{ijrs} w_{i,j} w_{r,s} dv = 0. \qquad (2.3.2)$$

In view of (2.3.1), we conclude that $w_{i,j} = 0$ almost everywhere in B and hence that the vector field $\underline{w}$ is almost everywhere a constant vector. In the displacement boundary value problem since $\underline{w} \in C^0(\bar{B})$ and vanishes on ∂B, it follows that $\underline{w} = \underline{0}$ in B. □

Note that instead of (2.3.1) a negative definite condition would serve equally well.

When d_{ijrs} are constants, we can prove a much stronger uniqueness theorem for the displacement problem.

Theorem 2.3.2. Suppose that d_{ijrs} are constants, and satisfy the strong ellipticity condition

$$d_{ijrs} t_i t_r v_j v_s \geq c t_i t_i v_j v_j, \qquad (2.3.3)$$

for some positive constant c and arbitrary vectors $\underline{t}, \underline{v}$. Then the displacement problem has at most one solution.

Proof. It is sufficient to show that $\underline{F} = \underline{0}$ on B and

$$\underline{w} = \underline{0} \text{ on } \partial B \qquad (2.3.4)$$

imply $\underline{w} = \underline{0}$ on B. Using the same notation and argument of the previous theorem, we obtain (2.3.2). Let us extend the definition of $\underline{w}$ from $\bar{B}$ to the whole space E^3 by defining

$$\underline{w} = \underline{0} \text{ on } E^3 - \bar{B}. \qquad (2.3.5)$$

Then, by (2.3.4) $\underline{w}$ is continuous and piecewise smooth on E^3, and $w_{i,j}$ has discontinuities only on ∂B. Thus w_i and $w_{i,j}$ are both absolutely and square integrable over E^3 and possess the three-dimensional Fourier transforms z_i and z_{ij}, respectively, given by

$$z_j(\underline{y}) = (2\pi)^{-3/2} \int_{E^3} e^{ix_s y_s} w_j \, dv,$$

$$_{mn}(\underline{y}) = (2\pi)^{-3/2} \int_{E^3} e^{ix_s y_s} w_{m,n} dv. \tag{2.3.6}$$

In view of (2.3.5), (2.3.6) and the divergence theorem, we obtain

$$z_{rs}(\underline{y}) = - iy_s z_r(\underline{y}). \tag{2.3.7}$$

A fundamental theorem of Fourier analysis (Goldberg [131]) yields

$$\int_{E^3} w_{i,j} w_{r,s} dv = \int_{E^3} \bar{z}_{ij} z_{rs} dv, \tag{2.3.8}$$

where $\bar{f}$ denotes the complex conjugate of f. By (2.3.7) and (2.3.8),

$$\int_B w_{i,j} w_{r,s} dv = \int_{E^3} x_j x_s \bar{z}_i z_r dv. \tag{2.3.9}$$

Since d_{ijrs} are constants, with the aid of (1.2.45), (2.3.3) and (2.3.9) we obtain

$$\int_B d_{ijrs} w_{i,j} w_{r,s} dv = d_{ijrs} \int_{E^3} w_{i,j} w_{r,s} dv =$$

$$= \text{Re} \int_{E^3} d_{ijrs} x_j x_s \bar{z}_i z_r dv \geq c \int_{E^3} x_s x_s \bar{z}_j z_j dv = \tag{2.3.10}$$

$$= c \int_B w_{j,s} w_{j,s} dv.$$

It follows from (2.3.2) and (2.3.10) that $w_{i,j}$ vanishes almost everywhere in B. The desired result follows from (2.3.4). □

We note that the theorem holds if $-d_{ijrs}$ is strongly elliptic. Theorem 2.3.2 follows from the work of Browder [39] and Morrey [265]. The proof given here employs the method of van Hove [182] (see also Knops and Payne [227]).

Hayes [169] established a uniqueness theorem for the displacement problem under a slightly weaker hypothesis on d_{ijrs}, which he calls moderate strong ellipticity. This is defined as

(α) $d_{ijrs}t_i t_r g_j g_s \geqq 0$ for all t_i, g_j;

(β) for $g_i \neq 0$, $d_{ijrs}t_i t_r g_j g_s = 0$ implies $t_i = 0$.

It is not difficult to see that moderate strong ellipticity and strong ellipticity are equivalent notions when the body is isotropic and unstressed.

Other results concerned with uniqueness for the displacement problem are given in [389], [167], [164], [170], [227].

We now give a uniqueness theorem for the mixed boundary value problem, established by Hayes [168]. We suppose that B_o is a bounded regular region, the portion of the boundary Σ_1 is star-shaped with respect to an origin located on Σ_2, i.e.

$$N_K X_K > 0 \text{ on } \Sigma_1, \tag{2.3.11}$$

while Σ_2 is planar, i.e.

$$N_K X_K = 0 \text{ on } \Sigma_2. \tag{2.3.12}$$

We assume that the primary state B is obtained from B_o by the homogeneous deformation $x_i = \lambda_{iK}X_K$ where λ_{iK} are given constants. We refer all quantities to the configuration B_o.

<u>Theorem 2.3.3.</u> Let B_o be a bounded region whose boundary consists of a planar part Σ_2 with the remainder Σ_1 being star-shaped with respect to an origin on Σ_2. Suppose that D_{iAjK} are constants, and satisfy the strong ellipticity condition

$$D_{iAjK}t_i t_j v_A v_K \geqq D_o t_i t_i v_A v_A, \tag{2.3.13}$$

for some positive constant D_o and arbitrary vectors t_i, v_A. Then the mixed boundary-value problem has at most one solution.

<u>Proof.</u> Let $\underline{u}$ be the difference of two solutions to the problem. Then, we have

$$D_{iAjK}u_{j,KA} = 0 \text{ on } B_o, \tag{2.3.14}$$

and

$$u_i = 0 \text{ on } \bar{\Sigma}_1, \quad D_{iAjK}u_{j,K}N_A = 0 \text{ on } \Sigma_2. \tag{2.3.15}$$

By (1.2.38) and (2.3.14),

$$\begin{aligned} 2D_{iAjK}u_{j,KA}u_{i,L}X_L &= 2(X_L D_{iAjK}u_{j,K}u_{i,L})_{,A} - \\ &- (X_L D_{iAjK}u_{i,A}u_{j,K})_{,L} + (D_{iAjK}u_{j,K}u_i)_{,A} = 0. \end{aligned} \tag{2.3.16}$$

The divergence theorem,(2.3.15) and (2.3.16) imply

$$\int_{\Sigma_1} (2X_L u_{i,L}D_{iAjK}u_{j,K}N_A - X_L N_L D_{iAjK}u_{i,A}u_{j,K})dA = 0. \tag{2.3.17}$$

Since $u_i = 0$ on $\bar{\Sigma}_1$, we have

$$u_{i,A} = a_i N_A \text{ on } \Sigma_1, \tag{2.3.18}$$

where

$$a_i a_i = u_{i,A}u_{i,A}. \tag{2.3.19}$$

It follows from (2.3.17) and (2.3.18) that

$$\int_{\Sigma_1} X_L N_L D_{iAjK}a_i a_j N_A N_K dA = 0. \tag{2.3.20}$$

By (2.3.11), (2.3.13) and (2.3.20) we obtain $a_i = 0$, and hence by (2.3.19),

$$u_{i,A} = 0 \text{ on } \Sigma_1. \tag{2.3.21}$$

Thus, we see that $u_i = 0$ and $u_{i,A} = 0$ on Σ_1. Moreover, we note that the strong ellipticity implies ellipticity so that the characteristic surfaces are imaginary. It then follows from Holmgren's theorem (see Knops and Payne

[227], p. 36) that $\underline{u} = \underline{0}$ on B_o. □

Counter-examples given by Hayes [167], [168] show that the strong ellipticity condition is not necessary for uniqueness of the boundary-value problems of superimposed infinitesimal deformations. Hayes [167] classified cases when uniqueness does and does not hold, correlating them with the signs of the squared wave speeds.

Other uniqueness results in the theory of infinitesimal elastic deformations superimposed on large are contained in the work of Knops and Payne [227].

2.4. Initial finite uniform extensions

We shall assume throughout this section that the body B_o is homogeneous and isotropic. We suppose that the primary state is obtained from B_o by the uniform finite extensions

$$x_1 = \lambda_1 X_1, \quad x_2 = \lambda_2 X_2, \quad x_3 = \lambda_3 X_3, \tag{2.4.1}$$

where λ_1, λ_2, λ_3 are the constant extension ratios. In this case, the basic equations have been discussed by Green and Zerna [140], Sect. 4.2. It follows from (1.1.12), (1.3.2), (1.3.4), (1.3.11) and (2.4.1) that

$$\begin{aligned}
&C_{11} = \lambda_1^2,\ C_{22} = \lambda_2^2,\ C_{33} = \lambda_3^2,\ C_{AB} = 0 (A \neq B),\\
&C_{11}^{(-1)} = \lambda_1^{-2},\ C_{22}^{(-1)} = \lambda_2^{-2},\ C_{33}^{(-1)} = \lambda_3^{-2},\ C_{AB}^{(-1)} = 0\ (A \neq B),\\
&I_1 = \lambda_1^2 + \lambda_2^2 + \lambda_3^2, \quad I_2 = \lambda_1^2\,\lambda_2^2 + \lambda_1^2\,\lambda_3^2 + \lambda_2^2\,\lambda_3^2,\\
&J = \lambda_1\,\lambda_2\,\lambda_3,\ b_{11} = \lambda_1^2,\ b_{22} = \lambda_2^2,\ b_{33} = \lambda_3^2, \qquad\qquad (2.4.2)\\
&b_{ij} = 0\ (i \neq j),\ B_{11} = \lambda_1^2\,\Lambda_{11} = (\lambda_2^2 + \lambda_3^2)\lambda_1^2,\\
&B_{22} = \lambda_2^2\,\Lambda_{22} = \lambda_2^2(\lambda_1^2 + \lambda_3^2),\ B_{33} = \lambda_3^2\,\Lambda_{33} =\\
&= \lambda_3^2(\lambda_1^2 + \lambda_2^2),\ B_{ij} = 0\ (i \neq j),
\end{aligned}$$

and from (1.3.7) we see that Φ, Ψ and p are constants. The stress components

T_{AB} and t_{ij} are given by (1.3.8) and (1.3.12) so that

$$T_{11} = \lambda_1 \lambda_2 \lambda_3[\Phi + (\lambda_2^2 + \lambda_3^2)\Psi + \lambda_1^{-2}p],$$

$$T_{22} = \lambda_1 \lambda_2 \lambda_3[\Phi + (\lambda_1^2 + \lambda_3^2)\Psi + \lambda_2^{-2}p],$$

$$T_{33} = \lambda_1 \lambda_2 \lambda_3[\Phi + (\lambda_1^2 + \lambda_2^2)\Psi + \lambda_3^{-2}p],$$

$$T_{AB} = 0(A \neq B), \tag{2.4.3}$$

$$t_{11} = \lambda_1^2\Phi + \lambda_1^2(\lambda_2^2 + \lambda_3^2)\Psi + p,$$

$$t_{22} = \lambda_2^2\Phi + \lambda_2^2(\lambda_1^2 + \lambda_3^2)\Psi + p,$$

$$t_{33} = \lambda_3^2\Phi + \lambda_3^2(\lambda_1^2 + \lambda_2^2)\Psi + p, \; t_{ij} = 0(i \neq j).$$

Since the stress tensor is constant, the equations of equilibrium are satisfied in the absence of body forces. By (1.3.18) and (2.4.2) we obtain

$$\alpha_{ij}^{(r)} = 0 \; (i \neq j), \; r = -1,0,1),$$

$$\alpha_{ii}^{(-1)} = 2p + 2\lambda_1^2\lambda_2^2\lambda_3^2[A_{13} \lambda_i^2 + A_{23} \lambda_i^2(\lambda_1^2 + \lambda_2^2 + \lambda_3^2 - \lambda_i^2) + A_{33} \lambda_1^2 \lambda_2^2 \lambda_3^2],$$

$$\alpha_{ii}^{(o)} = 2\lambda_i^2\Psi + 2[(A_{11} + I_1A_{12})\lambda_i^2 + (A_{12} + I_1A_{22})B_{ii} + (A_{13} + I_1A_{33})I_3], \tag{2.4.4}$$

$$\alpha_{ii}^{(1)} = -2[A_{12} \lambda_i^2 + A_{22} \lambda_i^2(\lambda_1^2 + \lambda_2^2 + \lambda_3^2 - \lambda_i^2) + A_{23} \lambda_1^2 \lambda_2^2 \lambda_3^2], \quad (i \text{ not summed}).$$

From (1.3.17) and (2.4.4) it follows that only non-zero coefficients C_{ijrs} are

C_{1111}, C_{1122}, C_{2222}, C_{1133}, C_{2233}, C_{3333}, C_{2323}, C_{1313}, C_{1212}.

Using (1.2.43) and (1.2.44) we may therefore write

$$
\begin{aligned}
s_{11} &= (c_{11}+t_{11})u_{1,1} + c_{12}u_{2,2} + c_{13}u_{3,3},\\
s_{22} &= c_{12}u_{1,1} + (c_{22}+t_{22})u_{2,2} + c_{23}u_{3,3},\\
s_{33} &= c_{13}u_{1,1} + c_{23}u_{2,2} + (c_{33} + t_{33})u_{3,3},\\
s_{23} &= c_{44}u_{2,3} + (c_{44} + t_{22})u_{3,2},\\
s_{32} &= (c_{44} + t_{33})u_{2,3} + c_{44}u_{3,2},\\
s_{31} &= (c_{55} + t_{33})u_{1,3} + c_{55}u_{3,1},\\
s_{13} &= c_{55}u_{1,3} + (c_{55} + t_{11})u_{3,1},\\
s_{12} &= c_{66}u_{1,2} + (c_{66} + t_{11})u_{2,1},\\
s_{21} &= (c_{66} + t_{22})u_{1,2} + c_{66}u_{2,1},
\end{aligned}
\tag{2.4.5}
$$

where

$$
\begin{aligned}
&c_{11} = C_{1111},\ c_{12} = C_{1122},\ c_{13} = C_{1133},\ c_{22} = C_{2222},\\
&c_{23} = C_{2233},\ c_{33} = C_{3333},\ c_{44} = C_{2323},\ c_{55} = C_{1313},\\
&c_{66} = C_{1212}.
\end{aligned}
\tag{2.4.6}
$$

It follows from (1.3.17), (2.4.2) and (2.4.4) that

$$
\begin{aligned}
c_{11} = {}& 2\lambda_1^4A_{11} + 2\lambda_1^4(\lambda_2^2 + \lambda_3^2)^2A_{22} + 2\lambda_1^4\lambda_2^4\lambda_3^4A_{33} +\\
&+ 4\lambda_1^4(\lambda_2^2 + \lambda_3^2)A_{12} + 4\lambda_1^4\lambda_2^2\lambda_3^2A_{13} + 4\lambda_1^4\lambda_2^2\lambda_3^2(\lambda_2^2 + \lambda_3^2)A_{23},
\end{aligned}
$$

$$c_{12} = 2p + 2\Psi\lambda_1^2\lambda_2^2 + 2\lambda_1^2\lambda_2^2 A_{11} + 2\lambda_1^2\lambda_2^2(\lambda_1^2 + \lambda_3^2)(\lambda_2^2 + \lambda_3^2)A_{22} +$$

$$+ 2\lambda_1^4\lambda_2^4\lambda_3^4 A_{33} + 2\lambda_1^2\lambda_2^2\lambda_3^2(2\lambda_1^2\lambda_2^2 + \lambda_2^2\lambda_3^2 + \lambda_1^2\lambda_3^2)A_{23} +$$

$$+ 2\lambda_1^2\lambda_2^2\lambda_3^2(\lambda_1^2 + \lambda_2^2)A_{13} + 2\lambda_1^2\lambda_2^2(\lambda_1^2 + \lambda_2^2 + \qquad (2.4.7)$$

$$+ 2\lambda_3^2)A_{12},$$

c_{22}, c_{33} being obtained from c_{11} by cyclic permutation of $\lambda_1,\lambda_2,\lambda_3$ and c_{23}, c_{13} being obtained from c_{12} by cyclic permutation of λ_1, λ_2, λ_3. Also,

$$c_{44} = -p-\lambda_2^2\,\lambda_3^2\Psi,\; c_{55} = -p-\lambda_1^2\lambda_3^2\Psi,\; c_{66} = -p-\lambda_1^2\lambda_2^2\Psi. \qquad (2.4.8)$$

We now consider a special case in which the extension ratios λ_1 and λ_2 are equal. By (2.4.3),

$$t_{11} = t_{22} = \lambda_1^2\Phi + \lambda_1^2(\lambda_1^2 + \lambda_3^2)\Psi + p,$$

$$t_{33} = \lambda_3^2\Phi + 2\lambda_1^2\,\lambda_3^2\Psi + p. \qquad (2.4.9)$$

It follows from (2.4.7) that

$$c_{11} = c_{22},\; c_{44} = c_{55},\; c_{13} = c_{23},\; 2c_{66} = c_{11} - c_{12}. \qquad (2.4.10)$$

The equations (2.4.5) become

$$s_{11} = a_{11}u_{1,1} + a_{12}u_{2,2} + a_{13}u_{3,3},$$

$$s_{22} = a_{12}u_{1,1} + a_{11}u_{2,2} + a_{13}u_{3,3},$$

$$s_{33} = a_{13}(u_{1,1} + u_{2,2}) + a_{33}u_{3,3}, \qquad (2.4.11)$$

$$s_{23} = a_{44}u_{2,3} + a_{45}u_{3,2},\; s_{32} = a_{54}u_{2,3} + a_{44}u_{3,2},$$

$$s_{31} = a_{54}u_{1,3} + a_{44}u_{3,1},\; s_{13} = a_{44}u_{1,3} + a_{45}u_{3,1},$$

$$s_{12} = a_{66}u_{1,2} + a_{65}u_{2,1},\; s_{21} = a_{65}u_{1,2} + a_{66}u_{2,1},$$

where

$$a_{11} = c_{11} + t_{11}, \ a_{12} = c_{12}, \ a_{13} = c_{13}, \ a_{33} = c_{33} + t_{33},$$

$$a_{44} = c_{44}, \ a_{45} = c_{44} + t_{11}, \ a_{54} = c_{44} + t_{33}, \tag{2.4.12}$$

$$2a_{66} = a_{11} - a_{12} - t_{11}, \ a_{65} = a_{66} + t_{11}.$$

Let $\mathcal{D}$ be the set of all admissible displacement fields on $\bar{B}$, and let $\underline{L}$ be the operator on $\mathcal{D}$ defined by

$$L_1\underline{u} = a_{11}u_{1,11} + \frac{1}{2}(a_{11}-d_{12})u_{1,22} + a_{54}u_{1,33} +$$

$$+ \frac{1}{2}(a_{11} + d_{12})u_{2,12} + (d_{13} + a_{45})u_{3,13},$$

$$L_2\underline{u} = \frac{1}{2}(a_{11} + d_{12})u_{1,12} + \frac{1}{2}(a_{11}-d_{12})u_{2,11} +$$

$$+ a_{11}u_{2,22} + a_{54}u_{2,33} + (d_{13} + a_{45})u_{3,23}, \tag{2.4.13}$$

$$L_3\underline{u} = (d_{13} + a_{45})(u_{1,13} + u_{2,23}) + a_{45}u_{3,\alpha\alpha} +$$

$$+ a_3u_{3,33},$$

where

$$d_{12} = a_{12} - t_{11}, \ d_{13} = a_{13} - t_{11}. \tag{2.4.14}$$

The equations of equilibrium (2.1.2) reduce to

$$L_i\underline{u} + \rho F_i = 0. \tag{2.4.15}$$

For the remainder of this section we assume that the body is incompressible. In this case $J = J^* = 1$ so that we have

$$\lambda_1 \lambda_2 \lambda_3 = 1, \quad u_{i,i} = 0. \tag{2.4.16}$$

It follows from (1.3.21), (1.3.22), (1.3.25) and (1.3.28) that

$$s_{11} = p' + d^o_{11}u_{1,1} + d^o_{12}u_{2,2} + d^o_{13}u_{3,3},$$
$$s_{22} = p' + d^o_{12}u_{1,1} + d^o_{22}u_{2,2} + d^o_{23}u_{3,3}, \qquad (2.4.17)$$
$$s_{33} = p' + d^o_{13}u_{1,1} + d^o_{23}u_{2,2} + d^o_{33}u_{3,3},$$

where

$$d^o_{11} = t_{11} - 2p + c^o_{11},\ d^o_{22} = t_{22} - 2p + c^o_{22},$$
$$d^o_{33} = t_{33} - 2p + c^o_{33},\ d^o_{ij} = c^o_{ij}\ (i \neq j),$$
$$c^o_{11} = 2\lambda_1^4[A^o_{11} + (\lambda_2^2 + \lambda_3^2)^2 A^o_{22} + 2(\lambda_2^2 + \lambda_3^2)A^o_{12}], \qquad (2.4.18)$$
$$c^o_{12} = 2\lambda_1^2\lambda_2^2[\Psi^o + A^o_{11} + (\lambda_1^2 + \lambda_3^2)(\lambda_2^2 + \lambda_3^2)A^o_{22} +$$
$$+ (\lambda_1^2 + \lambda_2^2 + 2\lambda_3^2)A^o_{12}],$$

c^o_{22}, c^o_{33} being obtained from c^o_{11} by cyclic permutation of $\lambda_1, \lambda_2, \lambda_3$, and c^o_{23}, c^o_{31} being obtained from c^o_{12} by cyclic permutation of $\lambda_1, \lambda_2, \lambda_3$. The remaining stress components are given by (2.4.5) and (2.4.8), with Ψ replaced by Ψ^o.

We now assume that $\lambda_1 = \lambda_2$. Then we obtain

$$s_{11} = p' + au_{1,1} + bu_{2,2},\ s_{22} = p' + bu_{1,1} + au_{2,2},$$
$$s_{33} = p' + cu_{3,3}, \qquad (2.4.19)$$

where

$$a = t_{11} - 2\lambda_1^2\lambda_3^2\Psi^o - 2p + 2\lambda_1^2(\lambda_1^2 - \lambda_3^2)Q,$$
$$b = 2\lambda_1^2(\lambda_1^2 - \lambda_3^2)(\Psi^o + Q),$$

$$c = t_{33} - 2\lambda_1^2\lambda_3^2\Psi^o - 2p + 2\lambda_3^2(\lambda_3^2 - \lambda_1^2)(A_{11}^o + $$
$$+ 2A_{22}^o\,\lambda_1^4 + 3A_{12}^o\,\lambda_1^2), \tag{2.4.20}$$
$$Q = A_{11}^o + \lambda_1^2(\lambda_1^2 + \lambda_3^2)A_{22}^o + (2\lambda_1^2 + \lambda_3^2)A_{12}^o.$$

The remaining components of stress are given by

$$s_{23} = a_{44}^o u_{2,3} + a_{45}^o u_{3,2}, \quad s_{32} = a_{54}^o u_{2,3} + a_{44}^o u_{3,2},$$
$$s_{31} = a_{54}^o u_{1,3} + a_{44}^o u_{3,1}, \quad s_{13} = a_{44}^o u_{1,3} + a_{45}^o u_{3,1}, \tag{2.4.21}$$
$$s_{12} = a_{66}^o u_{1,2} + a_{65}^o u_{2,1}, \quad s_{21} = a_{65}^o u_{1,2} + a_{66}^o u_{2,1},$$

where

$$a_{44}^o = c_{44}^o, \; a_{45}^o = c_{44}^o + t_{11}, \; a_{54}^o = c_{44}^o + t_{33},$$
$$2a_{66}^o = d_{11}^o - d_{12}^o - t_{11}, \; 2a_{65}^o = d_{11}^o - d_{12}^o + t_{11}. \tag{2.4.22}$$

We note that

$$a\text{-}b = t_{11} - 2p + c_{11}^o - c_{12}^o, \; 2a_{65}^o = 2t_{11} - 2p + c_{11}^o - c_{12}^o,$$
$$2a_{66}^o = c_{11}^o - c_{12}^o - 2p, \quad a\text{-}b\text{-}a_{66}^o = a_{65}^o. \tag{2.4.23}$$

2.5 Solutions of the displacement equations of equilibrium

In this section we study certain general solutions of the displacement equations of equilibrium governing the infinitesimal deformation superimposed on finite uniform extensions. We assume that the finite extension rates λ_1 and λ_2 are equal. We define

$$h_\alpha = (a_{11}m_\alpha - a_{54})/(d_{13} + a_{45}) =$$
$$= (d_{13} + a_{45})m_\alpha/(a_{33} - a_{45}m_\alpha), \tag{2.5.1}$$

where m_1, m_2 are the roots of the equation

$$a_{11}a_{45}m^2 + [(d_{13} + a_{45})^2 - a_{11}a_{33} - a_{45}a_{54}]m + a_{33}a_{54} = 0. \qquad (2.5.2)$$

Let

$$m_3 = 2a_{54} / (a_{11} - d_{12}), \qquad (2.5.3)$$

and let Δ_o be the two-dimensional Laplacian.

Theorem 2.5.1. Assume that m_α are real and $d_{13} + a_{45} \neq 0$, $a_{33} - a_{45}m_\alpha \neq 0$, $a_{11} \neq d_{12}$. Let

$$\begin{aligned} v_\alpha &= (g_1 + g_2)_{,\alpha} + e_{\alpha\beta 3}g_{3,\beta}, \\ v_3 &= (h_1g_1 + h_2g_2)_{,3}, \end{aligned} \qquad (2.5.4)$$

where g_i are class C^2 fields on B that satisfy

$$(\Delta_o + m_i \frac{\partial^2}{\partial x_3^2})g_i = 0 \text{ (i not summed)}. \qquad (2.5.5)$$

Then $\underline{v}$ satisfies the equations of equilibrium (2.4.15) with null body forces.

Proof. By (2.4.13), (2.5.3) and (2.5.4),

$$\begin{aligned} L_1\underline{v} &= \tfrac{1}{2}(a_{11}-d_{12})(\Delta_o + m_3 \frac{\partial^2}{\partial x_3^2})g_{3,2} + \{a_{11}\,\Delta_o + \\ &+ [a_{54} + h_1(d_{13} + a_{45})] \frac{\partial^2}{\partial x_3^2}\}\, g_{1,1} + \{a_{11}\,\Delta_o + \\ &+ [a_{54} + h_2(d_{13} + a_{45})] \frac{\partial^2}{\partial x_3^2}\}\, g_{2,1}, \\ L_2\underline{v} &= -\tfrac{1}{2}(a_{11} - d_{12})(\Delta_o + m_3 \frac{\partial^2}{\partial x_3^2})g_{3,1} + \\ &+ \{a_{11}\,\Delta_o + [a_{54} + h_1(d_{13} + a_{45})] \frac{\partial^2}{\partial x_3^2}\}g_{1,2} + \end{aligned} \qquad (2.5.6)$$

$$+ \{a_{11}\Delta_o + [a_{54} + h_2(d_{13} + a_{45})]\frac{\partial^2}{\partial x_3^2}\} g_{2,2},$$

$$L_3\underline{v} = \{(d_{13} + a_{45} + h_1 a_{45})\Delta_o g_1 + h_1 a_{33} g_{1,33} +$$

$$+ (d_{13} + a_{45} + h_2 a_{45})\Delta_o g_2 + h_2 a_{33} g_{2,33}\}_{,3}.$$

The relations (2.5.1), (2.5.5) and (2.5.6) imply the desired result. □

We introduce the notations

$$D_1 = a_{45}\Delta_o + a_{33}\frac{\partial^2}{\partial x_3^2},\quad D_2 = \frac{1}{2}(a_{11} - d_{12})\Delta_o + a_{54}\frac{\partial^2}{\partial x_3^2},$$

$$D_3 = \frac{1}{2}(a_{11} + d_{12})D_1 - (d_{13} + a_{45})^2\frac{\partial^2}{\partial x_3^2},$$

$$M = D_1 D_2 + \frac{1}{2}(a_{11} + d_{12})D_1\Delta_o - (d_{13} + a_{45})^2\Delta_o\frac{\partial^2}{\partial x_3^2} \tag{2.5.7}$$

$$Q = D_2 M.$$

A counterpart of the Boussinesq-Somigliana-Galerkin solution in the classical elastostatics (see, for example, Gurtin [153], Sect. 44) is given by

Theorem 2.5.2. Let

$$u_1 = (D_1 D_2 + D_3\frac{\partial^2}{\partial x_2^2})G_1 - D_3 G_{2,12} - (d_{13} + a_{45})G_{3,13},$$

$$u_2 = -D_3 G_{1,12} + (D_1 D_2 + D_3\frac{\partial^2}{\partial x_1^2})G_2 - (d_{13} + a_{45})G_{3,23}, \tag{2.5.8}$$

$$u_3 = -(d_{13} + a_{45})D_2 G_{\rho,\rho 3} + [D_2 + \frac{1}{2}(a_{11} + d_{12})\Delta_o]G_3,$$

where G_i are class C^6 fields on B that satisfy

$$QG_\alpha = -\rho F_\alpha,\quad MG_3 = -\rho F_3. \tag{2.5.9}$$

Then u_i satisfy the equations of equilibrium with the body forces F_j.

Proof. Clearly,

$$L_1\underline{u} = D_2 u_1 + \frac{1}{2}(a_{11} + d_{12})u_{\rho,\rho 1} + (d_{13} + a_{45})u_{3,31}. \tag{2.5.10}$$

By (2.5.7) and (2.5.10),

$$\begin{aligned} L_1\underline{u} &= D_1 D_2^2 G_1 + D_2 D_3(G_{1,22} - G_{2,21}) - (d_{13} + a_{45})D_2 G_{3,13} + \\ &+ \frac{1}{2}(a_{11}+d_{12})[D_1 D_2 G_{1,11} + D_1 D_2 G_{2,12} - (d_{13}+a_{45})\Delta_o G_{3,13}] + \\ &+ (d_{13}+a_{45})\{[D_2 + \frac{1}{2}(a_{11}+d_{12})\Delta_o]G_{3,13} - (d_{13}+a_{45})D_2 G_{\rho,\rho 133}\} = \\ &= D_1 D_2^2 G_1 + \frac{1}{2}(a_{11}+d_{12})D_1 D_2 \Delta_o G_1 - (d_{13}+a_{45})^2 D_2 \Delta_o G_{1,33} = -\rho F_1. \end{aligned}$$

Thus we conclude from (2.5.9) that the first equation of (2.4.15) is satisfied. In a similar manner we can prove that the other equations of (2.4.15) are satisfied. □

We note that

$$M = a_{11}a_{45}\Delta_o^2 + [a_{11}a_{33} + a_{45}a_{54} - (d_{13} + a_{45})^2]\Delta_o \frac{\partial^2}{\partial x_3^2} + \\ + a_{33}a_{54}\frac{\partial^4}{\partial x_3^4}.$$

Let m_1 and m_2 be the roots of the equation (2.5.2). If $a_{11} \neq d_{12}$, then

$$\begin{aligned} M &= a_{11}a_{45}(\Delta_o + m_1\frac{\partial^2}{\partial x_3^2})(\Delta_o + m_2\frac{\partial^2}{\partial x_3^2}), \\ Q &= (a_{11} - d_{12})(\Delta_o + m_3\frac{\partial^2}{\partial x_3^2})M. \end{aligned} \tag{2.5.11}$$

We now assume that the body is incompressible and that $a_{65}^o \neq 0$, $a_{54}^o \neq 0$. Let t_1 and t_2 be distinct roots of the equation

$$a_{45}^o t^2 + (2a_{44}^o - a - c)t + a_{54}^o = 0,$$

and

$$t_3 = a^0_{54} \,/\, a^0_{65}.$$

The proof of the next proposition can safely be omitted; it is completely analogous to the proof of Theorem 2.5.1.

Theorem 2.5.3. Let

$$\begin{aligned} w_\alpha &= (f_1 + f_2)_{,\alpha} + e_{\alpha\beta 3} f_{3,\beta}, \quad w_3 = (t_1 f_1 + t_2 f_2)_{,3}, \\ p' &= (at_1 - a^0_{44} t_1 - a^0_{54}) f_{1,33} + (at_2 - a^0_{44} t_2 - a^0_{54}) f_{2,33}, \end{aligned} \tag{2.5.12}$$

where f_i are class C^2 fields on B that satisfy

$$(\Delta_0 + t_i \frac{\partial^2}{\partial x_3^2}) f_i = 0 \quad \text{(i not summed)}. \tag{2.5.13}$$

Then $\underline{w}$ and p' satisfy the equations of equilibrium for incompressible bodies with null body forces.

The solutions (2.5.4) and (2.5.12) are due to Green and Zerna (see [140], Sect. 4.4). In the following section we will use the solution (2.5.12) to solve a special problem.

2.6 Indentation problems

In this section we suppose that the primary state B is obtained from the body B_0 by the uniform finite extensions (2.4.1) with $\lambda_1 = \lambda_2$. We assume that the body is incompressible, and that B occupies the region $x_3 \geqq 0$. The solution (2.5.12) is now used to solve a problem in which the stress component t_{33} is zero so that

$$p = -\lambda_3^2(\Phi + 2\Psi\lambda_1^2). \tag{2.6.1}$$

We assume that the secondary state is obtained from B by the indentation of the boundary $x_3 = 0$ by a rigid punch which has an axis of symmetry in the direction of the x_3-axis, and which is pressed normally against the plane

$x_3 = 0$. This problem was solved by Green and Zerna [140], Sect. 4.6. Let a be the radius of the circle of contact between the punch and the surface $x_3 = 0$ in the secondary configuration. We assume that the remainder of the boundary is free of applied forces and that the frictional forces are zero. The displacement component u_3 is prescribed on $x_3 = 0$, $r \leqq a$ where $r = (x_\alpha x_\alpha)^{1/2}$. Thus, for $x_3 = 0$, we have the conditions

$$\begin{aligned} u_3 &= g(r), \quad 0 \leqq r \leqq a, \\ s_{33} &= 0, \quad r > a, \\ s_{3\alpha} &= 0, \quad r \geqq 0, \end{aligned} \tag{2.6.2}$$

where g is a prescribed function which defines the shape of the punch. We assume that g is continuously differentiable.

We impose the further condition that s_{ij} and u_i vanish at infinity. It follows from (2.6.1), (2.4.21), (2.4.22) and (2.5.12) that

$$\begin{aligned} s_{3\alpha} &= c^o_{44}\{[(1+t_1)f_1 + (1+t_2)f_2]_{,\alpha 3} + e_{\alpha\beta 3}f_{3,\beta 3}\}, \\ s_{33} &= (dt_1 - c^o_{44})f_{1,33} + (dt_2 - c^o_{44})f_{2,33}, \end{aligned} \tag{2.6.3}$$

where $d = a + c - c^o_{44}$. The boundary conditions (2.6.2) can be satisfied if $f_3 = 0$ and

$$\begin{aligned} t_1 f_{1,3} + t_2 f_{2,3} &= g, \quad 0 \leqq r \leqq a, \\ (1+t_1)f_{1,3} + (1+t_2)f_{2,3} &= 0, \quad r > 0, \\ (dt_1 - c^o_{44})f_{1,33} + (dt_2 - c^o_{44})f_{2,33} &= 0, \quad r > a, \end{aligned} \tag{2.6.4}$$

for $x_3 = 0$. Let G be a class C^2 field on B that satisfies

$$\Delta G = 0. \tag{2.6.5}$$

Clearly, if we put

$$f_1 = \frac{t_1^{1/2}}{1+t_1} G(x_1,x_2,z_1), \quad f_2 = -\frac{t_2^{1/2}}{1+t_2} G(x_1,x_2,z_2), \tag{2.6.6}$$

where $z_\alpha = x_3 t_\alpha^{-1/2}$, then f_1 and f_2 satisfy the equations (2.5.13). Moreover, $s_{3\alpha} = 0$ for $x_3 = 0$. It follows from (2.5.12), (2.6.3) and (2.6.6) that

$$\begin{aligned} u_3 &= \frac{t_1}{1+t_1} \frac{\partial G}{\partial z_1} - \frac{t_2}{1+t_2} \frac{\partial G}{\partial z_2}, \\ s_{33} &= \frac{1}{1+t_1} (dt_1 - c_{44}^o) t_1^{-1/2} \frac{\partial^2 G}{\partial z_1^2} - \frac{1}{1+t_2} (dt_2 - c_{44}^o) t_2^{-1/2} \frac{\partial^2 G}{\partial z_2^2}. \end{aligned} \tag{2.6.7}$$

The conditions (2.6.4) reduce to

$$\begin{aligned} G_{,3}(x_1,x_2,0) &= \alpha g(r), \quad 0 \leqq r \leqq a, \\ G_{,33}(x_1,x_2,0) &= 0, \qquad r > a, \end{aligned} \tag{2.6.8}$$

where $\alpha = (1+t_1)(1+t_2)/(t_1-t_2)$. The problem is reduced to the boundary-value problem (2.6.5), (2.6.8), which frequently occurs in the classical theory of elasticity. We record that (Green and Zerna [140], Sect. 5.8)

$$G_{,3} = \frac{1}{2} \int_{-a}^{a} [r^2 + (x_3 + it)]^{-1/2} h(t)dt,$$

where h is given by

$$h = (2\alpha/\pi) \frac{d}{dt} \int_0^t (t^2-r^2)^{-1/2} rg(r)dr.$$

The detailed discussion of the problem can be made by arguments used in the classical theory (see Green and Zerna [140], Sect. 5.8).

In recent years various interesting results concerning indentation problems for initially stressed bodies have been established (see, for example, Babich [12] - [16], Dhaliwal, Singh and Rokne [105], Guz' and Babich [160], Guz' and Rudnitskij [162], Keer and Ballarini [219], Aleksandrov and Arutyunyan [3]).

2.7 Torsion of prestressed elastic cylinders

In this section we study the problem of torsion of a cylinder about an axis, after it has received a finite extension parallel to this axis. The result we give here is due to Green and Shield [132]. As remarked by Truesdell and Noll [356], Sect. 70, this is one of the important results in the equilibrium theory of infinitesimal deformations superimposed on large.

We suppose that the region B_0 refers to the interior of a right cylinder of length h with the open cross-section Σ_0 and the lateral boundary Π_0. We assume that the generic cross-section is a simply connected regular region. The rectangular Cartesian coordinate frame is supposed to be chosen in such a way that the X_3-axis coincides with the line of centroids of cross-sections, and X_1OX_2-plane contains one of the terminal cross-sections while the other is in the plane $X_3 = h$. We assume that the cylinder B_0 is homogeneous and isotropic. The body is first deformed into another cylinder B by a uniform finite extension along the X_3-axis. We let Π denote the lateral boundary of B, and designate by Σ the cross-section of B. We assume that the body forces are absent, so that the equations of equilibrium for the primary deformations are given by

$$T_{Ai,A} = 0 \text{ on } B_0. \tag{2.7.1}$$

The cylinder is assumed to be free from lateral loading. Then, the conditions on the lateral surface are

$$T_{Ai}N_A = 0 \text{ on } \Pi_0. \tag{2.7.2}$$

We suppose that the loading applied on the end located at $X_3 = 0$ is equivalent to a force $\underline{R}^0\ (0,0,R_3^0)$ and the moment $\underline{M}^0 = \underline{0}$. Thus, for $X_3 = 0$ we have the conditions

$$\int_{\Sigma_0} T_{3i}dA = -R_3^0\,\delta_{i3}, \quad \int_{\Sigma_0} e_{ijk}x_jT_{3k}dA = 0. \tag{2.7.3}$$

For the remainder of this chapter we shall denote the second Piola-Kirchhoff tensor by $T_{AM}^{(2)}$. We assume that the primary deformation is characterized by

$$x_\alpha = \lambda_1 X_\alpha, \quad x_3 = \lambda_3 X_3, \tag{2.7.4}$$

where λ_1 and λ_3 are unknown constants. Then, the only non-zero components t_{ij} are given by (2.4.9). It follows from (1.1.19) and (2.4.9) that the non-zero components of the first Piola-Kirchhoff stress tensor are given by

$$T_{11} = T_{22} = \lambda_1 \lambda_3 t_{11}, \quad T_{33} = \lambda_1^2 t_{33}, \tag{2.7.5}$$

so that the equations (2.7.1) are satisfied. The conditions (2.7.2) and (2.7.3) reduce to

$$t_{11} = 0, \quad \lambda_1^2 A_o t_{33} = - R_3^o, \tag{2.7.6}$$

where A_o is the area of the cross-section Σ_o.

Thus, the constants λ_1 and λ_3 are given by (2.7.6) and cannot be determined explicitly unless the particular form for the strain energy is known. For most bodies, however, it is probable that λ_1 will be greater than unity when λ_3 is less than unity.

Clearly, by (2.7.4) the cylinder B_0 is deformed into a right cylinder B, with the generators parallel to the X_3-axis. The end located at $X_3 = 0$ remains in this plane.

We assume that the secondary state B* is a configuration of equilibrium which is obtained from B by a torsion about the X_3-axis. We suppose that in the secondary deformation, B is free of lateral loading and that $f_i^* = 0$. The equations of equilibrium (2.1.2) become

$$s_{ji,j} = 0 \text{ on } B. \tag{2.7.7}$$

Since $t_{11} = t_{22} = 0$, the equations (2.4.11) reduce to

$$s_{11} = c_{11}u_{1,1} + c_{12}u_{2,2} + c_{13}u_{3,3},$$

$$s_{22} = c_{12}u_{1,1} + c_{11}u_{2,2} + c_{13}u_{3,3},$$

$$s_{33} = c_{13}(u_{1,1} + u_{2,2}) + (c_{33} + t_{33})u_{3,3},$$

$$s_{23} = c_{44}(u_{2,3} + u_{3,2}),$$

$$s_{32} = (c_{44} + t_{33})u_{2,3} + c_{44}u_{3,2}, \tag{2.7.8}$$

$$s_{31} = (c_{44} + t_{33})u_{1,3} + c_{44}u_{3,1},$$

$$s_{13} = c_{44}(u_{1,3} + u_{3,1}),$$

$$s_{12} = s_{21} = \frac{1}{2}(c_{11} - c_{12})(u_{1,2} + u_{2,1}).$$

The conditions on the lateral boundary are

$$s_{\alpha i}\, n_\alpha = 0 \text{ on } \Pi. \tag{2.7.9}$$

Note that

$$\int_{\Sigma_0} X_M\, dA = 0 \quad (M = 1,2). \tag{2.7.10}$$

We now suppose that, in the secondary deformation, the loading applied on the end located at $x_3 = 0$ is equivalent to the force $\underline{R}^*(0,0,R_3^o)$ and the moment $\underline{M}^*(0,0,M_3)$. Thus, for $x_3 = 0$ we have the conditions

$$\int_\Sigma s_{3\alpha}\, da = 0, \int_\Sigma (t_{33} + s_{33})da = -R_3^o,$$

$$\int_\Sigma (u_\alpha t_{33} + x_\alpha s_{33})da = 0, \tag{2.7.11}$$

$$\int_\Sigma e_{3\alpha\beta}x_\alpha\, s_{3\beta}\, da = -M_3.$$

The torsion problem consists in the determination of an admissible displacement field $\underline{u}$ on $\bar{B}$ that satisfies the equations (2.7.7), (2.7.8) and the conditions (2.7.9), (2.7.11), where M_3 is a given constant.

We seek the solution in the form

$$u_1 = -\tau x_2 x_3, \quad u_2 = \tau x_1 x_3, \quad u_3 = \tau\phi(x_1,x_2), \tag{2.7.12}$$

where ϕ is an unknown function and τ is an unknown constant. From (2.7.8) and (2.7.12) we obtain

$$s_{\alpha\beta} = s_{33} = 0, \quad s_{\alpha 3} = c_{44}\,\tau(\phi_{,\alpha} - e_{3\alpha\beta}x_\beta),$$
$$s_{3\alpha} = \tau[c_{44}(\phi_{,\alpha} - e_{3\alpha\beta}x_\beta) - e_{3\alpha\beta}x_\beta t_{33}]. \tag{2.7.13}$$

The equations (2.7.7) are satisfied provided

$$\phi_{,\alpha\alpha} = 0 \text{ on } \Sigma. \tag{2.7.14}$$

The boundary conditions (2.7.9) reduce to

$$\phi_{,\alpha}\, n_\alpha = e_{3\alpha\beta}x_\beta n_\alpha \text{ on } L, \tag{2.7.15}$$

where L is the boundary of Σ. Clearly, the boundary-value problem (2.7.14), (2.7.15) has solution. In what follows we assume that the function ϕ is known.

In view of (2.7.4)-(2.7.10) and (2.7.13) we obtain

$$\int_\Sigma s_{31}da = \int_\Sigma (s_{13} - \tau x_2 t_{33})da = \int_\Sigma s_{13}da =$$
$$= \int_\Sigma (x_1 s_{\alpha 3})_{,\alpha}\, da = \int_L x_1 s_{\alpha 3} n_\alpha\, ds = 0, \tag{2.7.16}$$
$$\int_\Sigma s_{32}da = 0, \int_\Sigma t_{33}da = -R_3^o.$$

Using (2.7.12), (2.7.13) and (2.7.16) we see that the conditions (2.7.11) reduce to

$$(c_{44}D + t_{33}I)\tau = -M_3, \tag{2.7.17}$$

where

$$D = \int_\Sigma (e_{3\alpha\beta}x_\alpha\, \phi_{,\beta} + x_\alpha x_\alpha)da, \; I = \int_\Sigma x_\alpha x_\alpha\, da. \tag{2.7.18}$$

It follows from (2.1.1) and (2.7.7) that

$$d_{ijrs}u_{i,j}u_{r,s} = s_{ji}u_{i,j} = (s_{ji}u_i)_{,j}.$$

By the divergence theorem and (2.7.9)

$$\int_B d_{ijrs}u_{i,j}u_{r,s}da = \int_{\Sigma_1 \cup \Sigma_2} s_{ji}n_j u_i da,$$

where Σ_1 and Σ_2 are the cross-sections of B located at $x_3 = 0$ and $x_3 = \lambda_3 h$, respectively. In view of (2.7.12), (2.7.13) and (2.7.16) we arrive at

$$\int_B d_{ijrs}u_{i,j}u_{r,s}dv = \lambda_3 h\tau^2(c_{44}D + t_{33}I).$$

Clearly, if d_{ijrs} is positive definite (or negative definite), then

$$c_{44}D + t_{33}I \neq 0. \tag{2.7.19}$$

The positive definiteness of the elasticities d_{ijrs} has been discussed by Knops and Wilkes [229] in connection with certain stability results.

For the remainder of this section we assume that (2.7.19) holds. Then the constant τ is determined by (2.7.17).

It follows from (2.4.8), (2.4.9) and (2.7.6) that

$$t_{33} = (\lambda_3^2 - \lambda_1^2)(\Phi + \lambda_1^2\Psi), \quad c_{44} = \lambda_1^2(\Phi + \lambda_1^2\Psi). \tag{2.7.20}$$

In view of (2.7.20), the relation (2.7.17) becomes

$$(\Phi + \lambda_1^2\Psi)[\lambda_3^2 I - \lambda_1^2(I - D)]\tau = - M_3. \tag{2.7.21}$$

We now introduce the function ϕ_o by

$$\phi_o(X_1,X_2) = \lambda_1^{-2}\phi(\lambda_1 X_1,\lambda_1 X_2).$$

The function ϕ_o satisfies the boundary-value problem

$$\phi_{o,\Lambda\Lambda} = 0 \text{ on } \Sigma_o, \quad \phi_{o,\Lambda}N_\Lambda = X_2N_1 - X_1N_2 \text{ on } L_o, \tag{2.7.22}$$

where L_o is the boundary of Σ_o. Thus, ϕ_o is the classical torsion function for a small torsion of the cylinder B_o. If we introduce the notations

$$I_o = \int_{\Sigma_o} X_\Lambda X_\Lambda \, dA, \quad D_o = \int_{\Sigma_o} (e_{3\Gamma\Lambda} X_\Gamma \phi_{o,\Lambda} + X_\Gamma X_\Gamma) dA,$$

then

$$I = \lambda_1^4 I_o, \quad D = \lambda_1^4 D_o,$$

and the relation (2.7.21) becomes

$$\lambda_1^4(\Phi + \lambda_1^2\Psi)[\lambda_3^2 I_o - (I_o - D_o)\lambda_1^2]\tau = - M_3. \tag{2.7.23}$$

Clearly, D_o is the classical torsional rigidity for the undeformed cylinder B_o, and I_o is the polar moment of inertia of the cross-section Σ_o.

From (2.7.6), (2.7.20) and (2.7.23) we obtain Green and Shield's universal relation

$$\tau R_3^o / M_3 = (\lambda_3^2 - \lambda_1^2) A_o \, \lambda_1^{-2} / [\lambda_3^2 I_o - (I_o - D_o)\lambda_1^2],$$

which connects the force required to produce the stretch λ_3 with the torsional modulus for a superimposed small twist. Since it is independent of the constitutive equation it may be used as a general test of the theory of isotropic elasticity (cf. Truesdell and Noll [356], Sect. 70).

2.8 The solution of a problem of Truesdell's

The statement of the torsion problem fails to characterize the solution uniquely. Let Q denote the set of all solutions of the torsion problem formulated in Section 2.7. In [358]-[360], C. Truesdell proposed the following problem: to define the functional $k(\cdot)$ on Q such that

$$(c_{44} D + t_{33} I) k(\underline{u}) = - M_3, \tag{2.8.1}$$

for each $\underline{u} \in Q$. A solution of Truesdell's problem in the absence of initial stresses has been established by Day [95]. Following Day, $k(\underline{u})$ is the generalized twist associated with $\underline{u}$. In [295], Podio-Guidugli solved Truesdell's problem rephrased for extension and pure bending. The flexure problem has been studied by Ieşan [197].

In this section we extend the solution due to Day to the theory of pre-stressed elastic cylinders.

Let $\underline{u}$ be an admissible displacement field on $\bar{B}$, and $s_{ij}(\underline{u})$ the stress field associated with $\underline{u}$, i.e.

$$s_{ji}(\underline{u}) = d_{ijrs}u_{r,s}.$$

Assume that the tensor field d_{ijrs} is positive definite. We call a vector field $\underline{u}$ an equilibrium displacement field if $\underline{u} \in C^1(\bar{B}) \cap C^2(B)$ and

$$(s_{ji}(\underline{u}))_{,j} = 0 \text{ on } B.$$

Let $\underline{s}(\underline{u})$ be the surface traction at regular points of ∂B corresponding to the stress tensor $s_{ji}(\underline{u})$, i.e. $s_i(\underline{u}) = s_{ji}(\underline{u})n_j$ on ∂B. The set of smooth vector fields over $\bar{B}$ can be made into a real vector space with the inner product

$$\langle \underline{u},\underline{v} \rangle = \int_B d_{ijrs}u_{i,j}u_{r,s}dv. \qquad (2.8.2)$$

This inner product generates the norm

$$\|\underline{u}\|_e^2 = \langle \underline{u},\underline{u} \rangle. \qquad (2.8.3)$$

It follows from (2.2.1), (2.2.3) and (2.2.5) that for any equilibrium displacement fields $\underline{u}$ and $\underline{v}$ one has

$$\langle \underline{u},\underline{v} \rangle = \int_{\partial B} \underline{s}(\underline{u})\cdot\underline{v}\ da, \qquad (2.8.4)$$

and

$$\int_{\partial B} \underline{s}(\underline{u}).\underline{v}\ da = \int_{\partial B} \underline{s}(\underline{v}).\underline{u}\ da. \qquad (2.8.5)$$

For any smooth vector field $\underline{u}$ we define the functionals

$$Y_i(\underline{u}) = -\int_\Sigma s_{3i}(\underline{u})da, \quad Z_\alpha(\underline{u}) = \int_\Sigma e_{3\beta\alpha}[x_\beta s_{33}(\underline{u}) + u_\beta t_{33}]da,$$

$$Z_3(\underline{u}) = \int_\Sigma e_{3\alpha\beta}x_\alpha s_{3\beta}(\underline{u})da, \quad x_3 = 0,$$

According to Section 2.7, by a solution of the torsion problem we mean any equilibrium vector field $\underline{u}$ that satisfies the conditions

$$\underline{s}(\underline{u}) = \underline{0} \text{ on } \Pi, \quad \underline{Y}(\underline{u}) = \underline{0},$$

$$Z_\alpha(\underline{u}) = 0, \quad Z_3(\underline{u}) = -M_3.$$

The set of all solutions of this problem was denoted by Q. Let H be the set of all equilibrium vector fields $\underline{u}$ that satisfy the conditions

$$\underline{s}(\underline{u}) = \underline{0} \text{ on } \Pi, \; [s_{33}(\underline{u})](x_1,x_2,0) = [s_{33}(\underline{u})](x_1,x_2,h),$$

$$\underline{Y}(\underline{u}) = \underline{0}, \quad Z_\alpha(\underline{u}) = 0, \quad Z_3(\underline{u}) = -M_3. \tag{2.8.6}$$

Clearly, H is a subclass of Q. Let $\underline{v}$ be the equilibrium vector field defined by

$$v_\alpha = e_{3\beta\alpha}x_\beta x_3, \quad v_3 = \phi(x_1,x_2), \tag{2.8.7}$$

where ϕ is the torsion function characterized by (2.7.14) and (2.7.15). Following Day [95], we define the real function

$$x \to \|\underline{u} - x\underline{v}\|_e^2,$$

where $\underline{u} \in H$ and $\underline{v}$ is given by (2.8.7). This function attains its minimum at

$$m(\underline{u}) = \langle\underline{u},\underline{v}\rangle / \|\underline{v}\|_e^2. \tag{2.8.8}$$

Let us prove that $k(\underline{u}) = m(\underline{u})$ for every $\underline{u} \in H$. In view of (2.8.4), (2.8.6) and (2.8.7) we find that

$$\langle \underline{u}, \underline{v} \rangle = \int_{\Sigma_1 \cup \Sigma_2} \underline{s}(\underline{u}).\underline{v}\, da = \lambda_3 h Z_3(\underline{u}). \tag{2.8.9}$$

Since

$$s_{3\alpha}(\underline{v}) = c_{44}(\phi_{,\alpha} - e_{3\alpha\beta}x_\beta) - e_{3\alpha\beta}x_\beta t_{33},$$

we arrive at

$$\|\underline{v}\|_e^2 = \lambda_3\, h(c_{44}D + t_{33}I), \tag{2.8.10}$$

where D and I are defined by (2.7.18). It follows from (2.8.8)-(2.8.10) that

$$Z_3(\underline{u}) = (c_{44}D + t_{33}I)m(\underline{u}). \tag{2.8.11}$$

By (2.8.1), (2.8.6) and (2.8.11) we find that $k(\underline{u}) = m(\underline{u})$ for each $\underline{u} \in H$. On the other hand, by (2.8.4), (2.8.5) and (2.8.7) we obtain

$$\langle \underline{u}, \underline{v} \rangle = T(\underline{u}), \tag{2.8.12}$$

where

$$T(\underline{u}) = \int_{\Sigma_2} u_\alpha[c_{44}(\phi_{,\alpha} - e_{3\alpha\beta}x_\beta) - e_{3\alpha\beta}x_\beta t_{33}]da -$$
$$- \int_{\Sigma_1} u_\alpha[c_{44}(\phi_{,\alpha} - e_{3\alpha\beta}x_\beta) - e_{3\alpha\beta}x_\beta t_{33}]da.$$

Thus, from (2.8.8), (2.8.10) and (2. 8.12) we conclude that

$$k(\underline{u}) = T(\underline{u})/[\lambda_3 h(c_{44}D + t_{33}I)],$$

for each $\underline{u} \in H$. This relation defines the generalized twist on the subclass H of solutions to the torsion problem.

2.9 Flexure, extension and bending

In Section 2.7 we have established a solution of torsion problem for an elastic cylinder which has been subjected to a large extension from a natural state. We now study the flexure, extension and bending. For convenience, the problem of flexure is treated separately. The flexure and bending solutions show that the large extension has a stiffening effect on a cylinder of isotropic material. In this section we assume that the X_3-axis is not restricted to coincide with the line of centroids of cross-sections. The resultant force and the resultant moment about 0 of the tractions acting on the end initially located at $X_3 = 0$, evaluated in the first deformation, are given by

$$R_i^o = -\int_\Sigma t_{3i} da, \quad M_i^o = -\int_\Sigma e_{ijk} x_j t_{3k} da \quad (x_3 = 0). \tag{2.9.1}$$

The corresponding quantities in the secondary deformation can be written in the form

$$R_i^* = -\int_\Sigma (t_{3i}+s_{3i}) da, \quad M_i^* = -\int_\Sigma e_{ijk}(x_j+u_j)(t_{3k}+s_{3k}) da \quad (x_3 = 0). \tag{2.9.2}$$

α) <u>Flexure</u>. When the secondary state B* is obtained from B by a flexure, then

$$R_\alpha^* - R_\alpha^o = R_\alpha, \quad R_3^* = R_3^o, \quad \underline{M}^* = \underline{M}^o, \tag{2.9.3}$$

where R_α are prescribed constants. In view of (2.4.3), (2.9.1) and (2.9.2), the conditions (2.9.3) take the form

$$\int_\Sigma s_{3\alpha}\, da = -R_\alpha, \tag{2.9.4}$$

$$\int_\Sigma s_{33} da = 0, \quad \int_\Sigma (u_\alpha t_{33} + x_\alpha s_{33}) da = 0, \tag{2.9.5}$$

$$\int_\Sigma e_{3\alpha\beta}\, x_\alpha s_{3\beta}\, da = 0, \tag{2.9.6}$$

for $x_3 = 0$. The flexure problem consists in the determination of an admissible displacement field $\underline{u}$ on $\bar{B}$ that satisfies the equations (2.7.7), (2.7.8) and the conditions (2.7.9), (2.9.4)-(2.9.6), where R_α are given constants. Following [197], we seek a solution of the flexure problem in the form

$$u_\alpha = b_\alpha x_3 + \tau e_{3\beta\alpha} x_\beta x_3, \quad u_3 = -b_\alpha x_\alpha + \tau\phi(x_1,x_2), \tag{2.9.7}$$

where ϕ is an unknown function, and b_α and τ are unknown constants. It follows from (2.7.8) and (2.9.7) that

$$\begin{aligned} s_{\alpha\beta} &= s_{33} = 0, \; s_{\alpha 3} = c_{44}\,\tau(\phi_{,\alpha} - e_{3\alpha\beta}x_\beta), \\ s_{3\alpha} &= s_{\alpha 3} + t_{33}(b_\alpha - \tau e_{3\alpha\beta}x_\beta). \end{aligned} \tag{2.9.8}$$

The equations (2.7.7) then reduce to the equation (2.7.4). The conditions (2.7.9) are satisfied if (2.7.15) holds. Clearly, ϕ is the classical warping function for the cross-section Σ in the strained reference state.

By (2.7.7) and (2.9.8)

$$\begin{aligned} s_{3\alpha} &= s_{\alpha 3} + x_\alpha s_{\beta 3,\beta} + t_{33}(b_\alpha - \tau e_{3\alpha\beta}x_\beta) = \\ &= (x_\alpha s_{\beta 3})_{,\beta} + t_{33}(b_\alpha - \tau e_{3\alpha\beta}x_\beta). \end{aligned} \tag{2.9.9}$$

Then (2.7.9), (2.9.9) and the divergence theorem imply that

$$\int_\Sigma s_{3\alpha}\, da = At_{33}(b_\alpha - \tau e_{3\alpha\beta}x^o_\beta), \tag{2.9.10}$$

where A is the area of the cross-section Σ, and x^o_α are given by

$$Ax^o_\alpha = \int_\Sigma x_\alpha\, da.$$

It follows from (2.9.4) and (2.9.10) that

$$b_\alpha - \tau e_{3\alpha\beta}x^o_\beta = -R_\alpha/At_{33}. \tag{2.9.11}$$

Using (2.9.7) and (2.9.8) we see that the conditions (2.9.5) are satisfied. The condition (2.9.6) reduces to

$$(c_{44}D + t_{33}I)\tau = - At_{33}(b_2x_1^o - b_1x_2^o), \tag{2.9.12}$$

where D and I are defined by (2.7.12).

As in Section 2.7, we now find that

$$\int_B d_{ijrs}u_{i,j}u_{r,s}dv = \lambda_3 h[At_{33}b_\alpha b_\alpha - e_{3\alpha\beta}x_\beta^o At_{33}\tau b_\alpha + \\ + \tau^2(c_{44}D + t_{33}I)]. \tag{2.9.13}$$

If d_{ijrs} is positive definite (or negative definite) then (2.7.19) holds. When (2.7.19) holds, the constants b_α and τ are determined by (2.9.11) and (2.9.12). It follows that (2.9.17) is a solution of the flexure problem.

The resultant force and the resultant moment about 0 of the tractions acting on the end located at $x_3 = \lambda_3 h$, evaluated in the first deformation, are given by

$$R_i' = \int_\Sigma t_{3i}da, \quad M_i' = \int_\Sigma e_{irs}x_r t_{3s}da \quad (x_3 = \lambda_3 h).$$

The corresponding quantities in the secondary deformation can be written in the form

$$R_i'' = \int_\Sigma (s_{3i}+t_{3i})da, \quad M_i'' = \int_\Sigma e_{irs}(x_r+u_r)(s_{3s}+t_{3s})da, \quad (x_3 = \lambda_3 h).$$

By (2.4.3), (2.9.8), (2.9.10), (2.9.11) and (2.9.12),

$$R_\alpha'' - R_\alpha' = \int_\Sigma s_{3\alpha}\, da = - R_\alpha, \quad R_3'' - R_3' = 0,$$

$$M_\alpha'' - M_\alpha' = e_{3\alpha\beta}\int_\Sigma (u_\beta t_{33} - \lambda_3 h s_{3\beta})da = 0,$$

$$M_3'' - M_3' = 0, \qquad (x_3 = \lambda_3 h).$$

Thus, we can see that the global conditions of equilibrium for the secondary state are satisfied.

Note that (2.9.1) implies

$$t_{33}A = - R_3^o. \tag{2.9.14}$$

If $x_\alpha^o = 0$ and $R_2 = 0$, then (2.9.11), (2.9.12) and (2.9.14) implies that $\tau = b_2 = 0$, and

$$b_1 = R_1/R_3^o. \tag{2.9.15}$$

This relation is independent of the response functions. In this case we obtain

$$u_1 = b_1 x_3, \quad u_2 = 0, \quad u_3 = -b_1 x_1. \tag{2.9.16}$$

The displacement field given by (2.9.16) represents a rigid rotation about the x_2-axis.

β) Extension and bending. We now assume that the secondary state B* is obtained from B by an extension and a bending, so that

$$R_\alpha^* = R_\alpha^o, \quad R_3^* - R_3^o = R_3, \quad M_\alpha^* - M_\alpha^o = M_\alpha, \quad M_3^* = M_3^o, \tag{2.9.17}$$

where R_3 and M_α are prescribed constants. By (2.4.3), (2.9.1) and (2.9.2), the conditions (2.9.17) reduce to

$$\int_\Sigma s_{3\alpha}\, da = 0, \tag{2.9.18}$$

$$\int_\Sigma s_{33} da = - R_3, \quad \int_\Sigma (x_\alpha\, s_{33} + u_\alpha t_{33}) da = e_{3\alpha\beta} M_\beta, \tag{2.9.19}$$

$$\int_\Sigma e_{3\alpha\beta}\, x_\alpha s_{3\beta}\, da = 0 \qquad (x_3 = 0). \tag{2.9.20}$$

The problem consists in the determination of an admissible displacement field $\underline{u}$ on $\bar{B}$ that satisfies the equations (2.7.7), (2.7.8) and the conditions (2.7.9), (2.9.18)-(2.9.20), where R_3 and M_α are given constants.

For the remainder of this section we assume that

$$c_{11} + c_{12} \neq 0. \tag{2.9.21}$$

We seek a solution of the problem in the form

$$u_\alpha = a_\alpha - a_3 \nu\, x_\alpha, \quad u_3 = a_3 x_3, \tag{2.9.22}$$

where

$$\nu = c_{13}/(c_{11} + c_{12}), \tag{2.9.23}$$

and a_i are unknown constants. It follows from (2.7.8), (2.9.22) and (2.9.23) that

$$s_{\alpha\beta} = s_{\alpha 3} = s_{3\alpha} = 0, \quad s_{33} = (c_{33}+t_{33}-2\nu c_{13})a_3. \tag{2.9.24}$$

Clearly, the equations (2.7.7) and the conditions (2.7.9), (2.9.18) and (2.9.20) are satisfied. The conditons (2.9.19) reduce to

$$\begin{aligned} &(c_{33} + t_{33} - 2\nu c_{13})Aa_3 = - R_3, \\ &At_{33}a_\alpha + [c_{33} - 2\,\nu\, c_{13} + (1 - \nu)t_{33}]x_\alpha^o \; Aa_3 = e_{3\alpha\beta}M_\beta. \end{aligned} \tag{2.9.25}$$

In this case we find that

$$\int_B d_{ijrs}u_{i,j}u_{r,s}dv = \lambda_3 hA(c_{33} + t_{33} - 2\nu c_{13}).$$

We assume that

$$c_{33} + t_{33} - 2\nu c_{13} \neq 0. \tag{2.9.26}$$

Then the constants a_i are determined by (2.9.25). We conclude that the displacement vector field $\underline{u}$ given by (2.9.22) is a solution of the extension and bending problem.

2.10 Torsion superimposed upon an infinitesimal extension

In this section we apply the theory of Section 1.5 to the problem of torsion of a rhombic crystal about a symmetry axis, after it has received an infinitesimal extension parallel to this axis. The results given here are due to Green [137].

We suppose that the region B_0 refers to the cylinder considered in Section 2.7. Moreover, the rectangular Cartesian coordinate frame is supposed to be chosen as in Section 2.7. We now assume that the body occupying B is homogeneous and has rhombic symmetry with respect to the rectangular Cartesian axes X_A. We suppose that the primary deformation is characterized by the equations (2.7.1), $(1.5.4)_1$, (1.5.5) and the boundary conditions (2.7.2) and (2.7.3). The non-zero constitutive coefficients are given by (1.5.12).

We seek a solution of this problem in the form

$$v_1 = -\alpha X_1, \quad v_2 = -\beta X_2, \quad v_3 = \gamma X_3, \tag{2.10.1}$$

where α, β and γ are unknown constants. By (1.5.5), (1.5.7) and (2.10.1),

$$\begin{gathered} \varepsilon_{11} = -\alpha, \ \varepsilon_{22} = -\beta, \ \varepsilon_{33} = \gamma, \ \varepsilon_{12} = \varepsilon_{23} = \varepsilon_{31} = 0, \\ J = 1 - \alpha - \beta + \gamma. \end{gathered} \tag{2.10.2}$$

It follows from (1.5.4), (1.5.12) and (2.10.1) that

$$\begin{aligned} T_{11}^{(2)} &= -\alpha c_{1111} - \beta c_{1122} + \gamma c_{1133}, \\ T_{22}^{(2)} &= -\alpha c_{1122} - \beta c_{2222} + \gamma c_{2233}, \\ T_{33}^{(2)} &= -\alpha c_{1133} - \beta c_{2233} + \gamma c_{3333}, \\ T_{12}^{(2)} &= T_{23}^{(2)} = T_{31}^{(2)} = 0, \quad T_{Ai} = T_{AM}^{(2)} \delta_{iM}. \end{aligned} \tag{2.10.3}$$

Clearly, the equations (2.7.1) are satisfied. The conditions (2.7.2) and (2.7.3) are satisfied if and only if

$$c_{1111}\alpha + c_{1122}\beta - c_{1133}\gamma = 0,$$

$$c_{1122}\alpha + c_{2222}\beta - c_{2233}\gamma = 0, \qquad (2.10.4)$$

$$c_{1133}\alpha + c_{2233}\beta - c_{3333}\gamma = R_3^o / A_o.$$

Thus, (α,β,γ) is a solution of the system (2.10.4). In what follows we assume that α, β and γ are given. Clearly, the cylinder is deformed into a right cylinder, B, with the generators parallel to the X_3-axis.

We assume that the secondary deformation is a torsion about the X_3-axis. The problem consists in determination of an admissible displacement field $\underline{u}$ on $\bar{B}$ that satisfies the equations (2.1.1), (2.7.7) and the conditions (2.7.9), (2.7.11), where M_3 is a given constant and d_{ijrs} are defined by (1.2.44), (1.2.30), (1.5.4), (1.5.12), (1.5.13), (2.10.1), (2.10.2). We seek a solution in the form

$$u_\alpha = \tau e_{3\beta\alpha} x_\beta x_3, \quad u_3 = \tau\phi(x_1,x_2), \qquad (2.10.5)$$

where ϕ is an unknown function and τ is an unknown constant. It follows from (1.1.19), (1.2.43), (1.2.44), (1.5.12), (1.5.13), (2.10.3)-(2.10.5) that

$$t_{ij} = \delta_{iA} T_{Aj}, \quad s_{\alpha\beta} = s_{33} = 0,$$

$$s_{31} = \tau[C_{1313}(\phi_{,1} - x_2) - t_{33}x_2],$$

$$s_{13} = \tau C_{1313}(\phi_{,1} - x_2), \qquad (2.10.6)$$

$$s_{32} = \tau[C_{2323}(\phi_{,2} + x_1) + t_{33}x_1],$$

$$s_{23} = \tau C_{2323}(\phi_{,2} + x_1),$$

where, retaining only linear terms in α, β and γ,

$$JC_{1313} = (1 - 2\alpha + 2\gamma)c_{1313} - \alpha c_{131311} - \beta c_{131322} + \gamma c_{131333},$$
$$JC_{2323} = (1 - 2\beta + 2\gamma)c_{2323} - \alpha c_{232311} - \beta c_{232322} + \gamma c_{232333}.$$

We assume that $C_{1313}C_{2323} > 0$. In view of (2.10.6), the equations (2.7.7) reduce to

$$C_{1313}\,\phi_{,11} + C_{2323}\,\phi_{,22} = 0 \text{ on } \Sigma. \tag{2.10.7}$$

The conditions (2.7.9) are satisfied if and only if

$$C_{1313}\phi_{,1}n_1 + C_{2323}\phi_{,2}n_2 = C_{1313}x_2n_1 - C_{2323}x_1n_2 \text{ on } L. \tag{2.19.8}$$

Clearly,

$$\int_L (C_{1313}x_2n_1 - C_{2323}x_1n_2)ds = 0,$$

so that the necessary and sufficient condition for the existence of a solution of the boundary value problem (2.10.7), (2.10.8) is satisfied.

It follows from (2.10.3)-(2.10.6) and (2.7.16) that the conditions (2.7.11) reduce to

$$(D^o + t_{33}I)\tau = -M_3, \tag{2.10.9}$$

where I is given by (2.7.18) and

$$D^o = \int_\Sigma [C_{2323}x_1(\phi_{,2} + x_1) - C_{1313}x_2(\phi_{,1} - x_2)]da.$$

If $D^o + t_{33}I \neq 0$, then the constant τ is determined by (2.10.9).

2.11. The plane problem

The plane problem for initially stressed bodies has been studied in various papers (see, for example, Guz' [159] and Manivachakan [250]). In this section we study the generalized plane strain problem for a homogeneously prestressed body. First, we derive fundamental solutions and establish relations of Somigliana type. Then, by using the method of potential (see Kupradze [237], Basheleishvili [19], Burchuladze [44]), the boundary-value problems are reduced to singular integral equations for which Fredholm's basic theorems are valid. The existence theorems for interior and exterior problems are derived. The generalized plane strain problem in the classical theory of elastostatics has been studied in various papers (see Lekhnitskii [243]). The results we give here are due to Ieşan [187]. We assume that the tensor d_{ijrs} is constant and positive definite. We suppose that the region B refers to the interior of a right cylinder with generators parallel to the x_3-axis and the open cross-section Σ. We assume that Σ is a simply connected regular plane region bounded by the closed Liapunov curve L.

Following Lekhnitskii [243], we define the state of generalized plane strain parallel to the plane x_1Ox_2, to be that state in which u_i depend only on x_1 and x_2. This restriction, in conjunction with the equations (2.1.1), implies that s_{ij} depend only on x_1 and x_2. The equilibrium equations (2.1.3) take the form

$$d_{i\alpha r\beta}\, u_{r,\alpha\beta} + \rho F_i = 0 \text{ on } \Sigma, \tag{2.11.1}$$

from which it follows that the functions F_i must be independent of x_3. The surface tractions acting at a point x on L are given by

$$s_i = s_{\alpha i} n_\alpha, \tag{2.11.2}$$

where $n_\alpha = \cos(\underline{n}_x, x_\alpha)$ and $\underline{n}_x$ is the unit vector of the outward normal to L at x.

α) Fundamental solutions. In [384], Woo and Shield have established the fundamental solutions for an incompressible homogeneous and isotropic body in finite biaxial extension. The solutions are analogous to those for an unstressed transversely isotropic body.

The equations (2.11.1) can be written in the form

$$A(\frac{\partial}{\partial x})u = f, \tag{2.11.3}$$

where

$$A(\frac{\partial}{\partial x}) = L_{\alpha\beta}\frac{\partial^2}{\partial x_\alpha \partial x_\beta}, \quad L_{\alpha\beta} = (L_{\alpha\beta ij}), \quad L_{\alpha\beta ij} = d_{i\alpha j\beta},$$

$$u = (u_1, u_2, u_3), \quad f = -\rho(F_1, F_2, F_3).$$

Throughout this section, a vector $v = (v_1, v_2, v_3)$ shall be also considered as a column-matrix.

Let

$$u = G(\frac{\partial}{\partial x})\phi \ , \ \phi = (\phi_1, \phi_2, \phi_3), \tag{2.11.4}$$

where ϕ_i are class C^6 fields, and

$$A(\frac{\partial}{\partial x})G(\frac{\partial}{\partial x}) = G(\frac{\partial}{\partial x})A(\frac{\partial}{\partial x}) = I \det A(\frac{\partial}{\partial x}). \tag{2.11.5}$$

Here, $I = (\delta_{ij})$. If we let

$$G(\frac{\partial}{\partial x}) = G_{\alpha\beta\kappa\rho} \frac{\partial^4}{\partial x_\alpha \ \partial x_\beta \ \partial x_\kappa \ \partial x_\rho} , \tag{2.11.6}$$

$$G_{\alpha\beta\kappa\rho} = (G_{ij\alpha\beta\kappa\rho}),$$

then

$$L_{\alpha\beta kj} \ G_{js\kappa\rho\mu\eta} \ \frac{\partial^6}{\partial x_\alpha \partial x_\beta \partial x_\kappa \partial x_\rho \partial x_\mu \partial x_\eta} = \delta_{ks} \det A(\frac{\partial}{\partial x}). \tag{2.11.7}$$

It follows from (2.11.3), (2.11.4) and (2.11.5) that

$$[\det A(\frac{\partial}{\partial x})]\phi_i = - \rho F_i. \tag{2.11.8}$$

Clearly, (2.11.4) is a solution of Galerkin type. We use this solution to determine the fundamental solutions of the equations (2.11.1). If $\rho F_i = \delta_{ip}\delta(x-y)$, where $\delta(\cdot)$ is the Dirac delta and $y \in \Sigma$ is a fixed point, then we take $\phi_i = \delta_{ip} e(x,y)$. Clearly, e satisfies the equation

$$[\det A(\frac{\partial}{\partial x})]e = - \delta(x-y). \tag{2.11.9}$$

In this case, from (2.11.4) we obtain the displacements $u_i^{(p)}(x,y)$,

$$u_i^{(p)}(x,y) = G_{ip\alpha\beta\kappa\rho} \ \frac{\partial^4 \ e(x,y)}{\partial x_\alpha \partial x_\beta \partial x_\kappa \partial x_\rho} . \tag{2.11.10}$$

The characteristic equation corresponding to the elliptic equation (2.11.9) is

$$L_{\alpha\beta 1j}\, G_{j1\kappa\rho\mu\eta}\, \alpha^{12-(\alpha+\beta+\kappa+\rho+\mu+\eta)} = 0. \tag{2.11.11}$$

The roots of this equation have the form

$$\alpha_s = p_s + iq_s,\ q_s > 0\ (s = 1,3,5),\ i = \sqrt{-1}\ ,\ \alpha_{s+1} = \bar{\alpha}_s.$$

We assume that $\alpha_1 \neq \alpha_3 \neq \alpha_5 \neq \alpha_1$. The other cases can be treated in a similar way. The function e is given by (cf. Levi [244])

$$e(x,y) = ia \sum_{j=1}^{6} (-1)^j d_j\, \sigma_j^4 \ln \sigma_j, \tag{2.11.12}$$

where

$$\sigma_j(x,y) = (x_1-y_1)\alpha_j + x_2 - y_2,$$

$$(j = 1,2,\ldots,6), \tag{2.11.13}$$

$$a^{-1} = 4!2\pi \det L_{11},$$

and d_j is the cofactor of α_j^5 from the determinant

$$d = \begin{vmatrix} \alpha_1^5 & \alpha_1^4 & \ldots & \alpha_1 & 1 \\ \alpha_2^5 & \alpha_2^4 & \ldots & \alpha_2 & 1 \\ \cdot & \cdot & \cdot & \cdot & \cdot \\ \alpha_6^5 & \alpha_6^4 & \ldots & \alpha_6 & 1 \end{vmatrix},$$

divided by d. Clearly, e is a real function. We have

$$\frac{\partial^4 e}{\partial x_\gamma \partial x_\delta \partial x_\eta \partial x_\rho} = ia \sum_{j=1}^{6} (-1)^j d_j \, \alpha_j^{8-(\gamma+\delta+\eta+\rho)} (4! \ln \sigma_j + 50),$$

$$\frac{\partial^5 e}{\partial x_\beta \partial x_\gamma \partial x_\delta \partial x_\delta \partial x_\rho} = 4! ia \sum_{j=1}^{6} (-1)^j \frac{1}{\sigma_j} d_j \alpha_j^{8-(\beta+\gamma+\delta+\eta+\rho)}. \tag{2.11.14}$$

It follows from (2.11.10) and (2.11.14) that

$$u_i^{(p)}(x,y) = \sum_{j=1}^{6} H_{ipj} \ln \sigma_j(x,y) + H_{ip}, \tag{2.11.15}$$

where

$$H_{rsj} = (-1)^j 4! \, iad_j \sum_{\alpha,\beta,\kappa,\rho=1}^{2} G_{rs\alpha\beta\kappa\rho} \, \alpha_j^{8-(\alpha+\beta+\kappa+\rho)},$$

$$H_{rs} = 50ia \sum_{j=1}^{6} \sum_{\alpha,\beta,\kappa,\rho=1}^{2} (-1)^j d_j G_{rs\alpha\beta\kappa\rho} \, \alpha_j^{8-(\alpha+\beta+\kappa+\rho)}. \tag{2.11.16}$$

β) Relations of Somigliana type. Let us consider two external data systems and the corresponding displacement fields u_i' and u_i''. It follows from (2.11.1), (2.11.2) and the divergence theorem that

$$\int_L s_i' u_i'' ds + \int_\Sigma \rho F_i' u_i'' da = 2 \int_\Sigma U(u',u'') da, \tag{2.11.17}$$

where

$$2U(u',u'') = d_{i\alpha j\beta} \, u'_{i,\alpha} \, u''_{j,\beta}, \quad s_i' = d_{i\alpha r\beta} \, u'_{r,\beta} \, n_\alpha. \tag{2.11.18}$$

In the view of (1.2.45) we obtain the reciprocity relation

$$\int_L s_i' u_i'' ds + \int_\Sigma \rho F_i' u_i'' da = \int_L s_i'' u_i' ds + \int_\Sigma \rho F_i'' u_i' da. \tag{2.11.19}$$

If we introduce the notations

$$u = (u_1', u_2', u_3'), \quad v = (u_1'', u_2'', u_3''),$$
$$s(u) = (s_1', s_2', s_3'), \ s(v) = (s_1'', s_2'', s_3''),$$

then the relation (2.11.17) can be written in the form

$$\int_L vs(u)ds - \int_\Sigma vA(\frac{\partial}{\partial x})uda = 2\int_\Sigma U(u,v)da. \tag{2.11.20}$$

This relation can be used to obtain a uniqueness result. Let us consider the boundary condition

$$s(u) = \tilde{s} \text{ on } L. \tag{2.11.21}$$

Let u^0 be a solution of the boundary-value problem (2.11.3) and (2.11.21) corresponding to $f = 0$ and $\tilde{s} = 0$. Since $U(u,u)$ is positive definite, it follows from (2.11.18) and (2.11.20) that

$$u^0 = (a_1,a_2,a_3), \tag{2.11.22}$$

where a_i are arbitrary constants. The relation (2.11.19) can be written in the form

$$\int_\Sigma [uA(\frac{\partial}{\partial x})v - vA(\frac{\partial}{\partial x})u]da = \int_L [us(v) - vs(u)]ds. \tag{2.11.23}$$

Let us introduce the matricial differential operator

$$H(\frac{\partial}{\partial x}, n_x) = (d_{i\alpha j\beta}n_\alpha \frac{\partial}{\partial x_\beta}). \tag{2.11.24}$$

It follows that

$$s(u) = H(\frac{\partial}{\partial x}, n_x)u. \tag{2.11.25}$$

We introduce the matrix of fundamental solutions of the system (2.11.3)

$$\Gamma(x,y) = (\Gamma_{ij}(x,y)), \ \Gamma_{ij} = u_i^{(j)}, \tag{2.11.26}$$

and denote by $\Gamma^{(k)}$ $(k = 1,2,3)$ the columns of the matrix Γ. By (2.11.15),

$$\Gamma = \sum_{j=1}^{6} M_j \ln \sigma_j + N, \tag{2.11.27}$$

where $M_j = (H_{rsj})$, $N = (H_{rs})$.

It follows from (2.11.23) that

$$u_k(y) = \int_L [\Gamma^{(k)}(x,y)s(u)-us^{(x)}(\Gamma^{(k)}(x,y))]ds_x - \\ - \int_\Sigma \Gamma^{(k)}(x,y)Auda, \tag{2.11.28}$$

where

$$s^{(x)}(\Gamma^{(k)}(x,y)) = H(\frac{\partial}{\partial x}, n_x)\Gamma^{(k)}(x,y). \tag{2.11.29}$$

We introduce the notations

$$s^{(x)}(\Gamma^{(k)}) = (T_1^{(x)}\Gamma^{(k)}, T_2^{(x)}\Gamma^{(k)}, T_3^{(x)}\Gamma^{(k)}).$$

Clearly,

$$T_p^{(x)} \Gamma^{(m)} = d_{p\alpha s\beta} n\alpha \Gamma_{sm,\beta}.$$

In view of (2.11.15) we obtain

$$T_p^{(x)} \Gamma^{(m)} = \sum_{j=1}^{6} d_{p\rho s\beta} H_{smj} \frac{1}{\sigma_j} \alpha_j^{2-\beta} n_\rho. \tag{2.11.30}$$

From (2.11.11) we get

$$d_{12j\beta} G_{j1\kappa\rho\mu\eta} \alpha^{10-(\beta+\kappa+\rho+\mu+\eta)} = -\alpha_j d_{11j\beta} G_{j1\kappa\rho\mu\eta} \alpha^{10-(\beta+\kappa+\rho+\mu+\eta)}.$$

Thus, we find that

$$T_p^{(x)} \Gamma^{(m)}(x,y) = \sum_{j=1}^{6} (-1)^j R_{pmj} d_j (n_1 - \alpha_j n_2) \frac{1}{\sigma_j}, \tag{2.11.31}$$

where

$$R_{pmj} = 4!ia \sum_{\beta,\gamma,\delta,\eta,\kappa=1}^{2} d_{p1s\beta} G_{sm\gamma\delta\eta\kappa} \alpha_j^{10-(\beta+\gamma+\delta+\eta+\kappa)}.$$

It is a simple matter to verify that

$$\frac{1}{\sigma_j}[\cos(n_x,x_1) - \alpha_j\cos(n_x,x_2)] = \frac{\partial}{\partial s_x}\ln\sigma_j,$$

$$\ln\sigma_j = \ln(\sigma_j/\sigma) - \frac{1}{2}\ln(\bar{\sigma}/\sigma) + \ln r, \tag{2.11.32}$$

$$\sigma = i(x_1-y_1) + x_2 - y_2,\ r^2 = (x_1-y_1)^2 + (x_2-y_2)^2,$$

$$\frac{\partial}{\partial s_x}\ln(\sigma_j/\sigma) = \frac{(i-\alpha_j)r^2}{\sigma\sigma_j}\ \frac{\partial}{\partial n_x}\ln r.$$

It follows from (2.11.29)-(2.11.32) that

$$H(\frac{\partial}{\partial x},n_x)\Gamma(x,y) = K\frac{\partial\ln r}{\partial s_x} + \Pi(x,y), \tag{2.11.33}$$

where

$$K = (K_{ij}),\ \Pi = (\Pi_{ij}),\ K_{pm} = \sum_{j=1}^{6}(-1)^j R_{pmj}d_j,$$

$$\Pi_{pm} = \sum_{j=1}^{6}(-1)^j R_{pmj}d_j[\frac{(i-\alpha_j)r^2}{\sigma\sigma_j} - i]\frac{\partial\ln r}{\partial n_x}. \tag{2.11.34}$$

If u satisfies the equation

$$A(\frac{\partial}{\partial x})u = 0, \tag{2.11.35}$$

then from (2.11.28) we obtain

$$u(x) = \int_L [\Gamma(x,y)\,H(\frac{\partial}{\partial y},n_y)u(y) - \Lambda(x,y)u(y)]ds_y, \tag{2.11.36}$$

where

$$\Lambda(x,y) = [H(\frac{\bar{\partial}}{\partial y},n_y)\,\Gamma(y,x)]^*. \tag{2.11.37}$$

We write S^* for the transpose of S.

Let Σ_e be the complementary of $\Sigma + L$ to the entire plane. Let x_o be a fixed point, $x_0 \in \Sigma$, and let $R = |x-x_0|$ be the distance between the points x and x_0. If the displacement fields satisfy

$$u_i = O(1), \quad u_{i,\alpha} = o(1/R), \tag{2.11.38}$$

then the reciprocity relation holds for the region Σ_e. By a regular displacement field on Σ we mean a vector field $\underline{u}$ such that $\underline{u} \in C^2(\Sigma) \cap C^1(\bar{\Sigma})$. A vector field $\underline{u}$ is a regular displacement field on Σ_e provided: (i) $\underline{u} \in C^2(\Sigma_e) \cap C^1(\bar{\Sigma}_e)$; (ii) $\underline{u}$ satisfies the conditions (2.11.38).

As in the classical theory (see also Kupradze [237]) we have

$$\int_L \Lambda(x,y)ds_y = -I\eta(x),$$

where $\eta(x) = 1$ if $x \in \Sigma$, $\eta(x) = 0$ if $x \in \Sigma_e$ and $\eta(x) = 1/2$ if $x \in L$.

γ) Reduction of the boundary-value problems to integral equations.

Let $\psi = (\psi_1, \psi_2, \psi_3)$ be a vector field on L whose components satisfy Hölder's condition. Then the vector ψ is said to be of class H. We shall now consider some basic boundary-value problems in the absence of body forces.

We introduce the potential of a single layer

$$V(x;\psi) = \int_L \Gamma(x,y)\psi(y)ds_y. \tag{2.11.39}$$

and the potential of a double layer

$$W(x;\psi) = \int_L \Lambda(x,y)\psi(y)ds_y. \tag{2.11.40}$$

Following the classical theory (see [19], [44], [243]) we can prove

Theorem 2.11.1. The potential of a single layer is continuous throughout.

Theorem 2.11.2. The potential of a double layer tends to finite limits when the point x tends to $z \in L$, both from within and from without, and these limits are respectively equal to

$$W_i(z;\psi) = -\frac{1}{2}\psi(z) + \int_L \Lambda(z,y)\psi(y)ds_y,$$

$$W_e(z;\psi) = \frac{1}{2}\psi(z) + \int_L \Lambda(z,y)\psi(y)ds_y.$$

Theorem 2.11.3. $H(\frac{\partial}{\partial x}, n_x)V(x;\psi)$ tends to finite limits when the point x tends to the boundary point $z \in L$, both from within and from without, and these limits are respectively equal to

$$[H(\frac{\partial}{\partial z}, n_z)V(z;\psi)]_i = \frac{1}{2}\psi(z) + \int_L [H(\frac{\partial}{\partial z}, n_z)\Gamma(z,y)]\psi(y)ds_y,$$

$$[H(\frac{\partial}{\partial z}, n_z)V(z;\psi)]_e = -\frac{1}{2}\psi(z) + \int_L [H(\frac{\partial}{\partial z}, n_z)\Gamma(z,y)]\psi(y)ds_y.$$

Moreover, we have

$$A(\frac{\partial}{\partial x})V(x;\psi) = 0, \quad A(\frac{\partial}{\partial x})W(x;\psi) = 0 \quad x \bar{\in} L.$$

In view of (2.11.27) and (2.11.30),

$$W = O(R^{-1}), \quad W_{,\alpha} = O(R^{-2}). \qquad (2.11.41)$$

Moreover, if

$$\int_L \psi(y)ds_y = 0, \qquad (2.11.42)$$

then

$$V = O(R^{-1}), \quad V_{,\alpha} = O(R^{-2}). \qquad (2.11.43)$$

We consider the equation (2.11.35) and the following boundary-value problems.

Interior problems. To find the solution regular in Σ of the equation (2.11.35) satisfying one of the conditions

(I_1) $\lim_{x \to z} u(x) = g(z),$

(I_2) $\lim_{x \to z} H(\frac{\partial}{\partial x}, n_x)u(x) = g(z),$

where $x \in \Sigma$, $z \in L$ and g is a given vector field of class H.

Exterior problems. To find the solution regular in Σ_e of the equation (2.11.35) satisfying one of the conditions

(E_1) $\lim_{x \to z} u(x) = g(z),$

(E_2) $\lim_{x \to z} H(\frac{\partial}{\partial x}, n_x)u(x) = g(z),$

where $x \in \Sigma_e$, $z \in L$ and g is a given vector field of class H.

The homogeneous problems corresponding to problems (I_1), (I_2), (E_1) and (E_2) will be denoted by (I_1^0), (I_2^0), (E_1^0) and (E_2^0), respectively. It follows from (2.11.20) the following uniqueness results.

Theorem 2.11.4. The solution regular in Σ of the problem (I_1^0) is identically zero. The solution regular in Σ of the problem (I_2^0) is given by (2.11.22). The solutions regular in Σ_e of the problems (E_1^0) and (E_2^0) are identically zero.

We seek the solutions of the problems (I_1) and (E_1) in the form of a double layer potential and the solutions of the problems (I_2) and (E_2) in the form of a single layer potential. We obtain for the unknown density the following singular integral equations

(I_1) $-\frac{1}{2}\psi(z) + \int_L \Lambda(z,y)\psi(y)ds_y = g(z),$

(I_2) $\frac{1}{2}\psi(z) + \int_L \Lambda^*(y,z)\psi(y)ds_y = g(z),$

(E_1) $\frac{1}{2}\psi(z) + \int_L \Lambda(z,y)\psi(y)ds_y = g(z),$

(E_2) $-\frac{1}{2}\psi(z) + \int_L \Lambda^*(z,y)\psi(y)ds_y = g(z).$

The homogeneous equations corresponding to equations (I_α) and (E_α) will be denoted by (I_α^o) and (E_α^o), respectively. The equations (I_1) and (E_2), (I_2) and (E_1) are pairwise mutually associate equations, as it is obvious from the form of their kernels. Using (2.11.33), (2.11.37) and the relation

$$\frac{\partial \ln r}{\partial s_y} ds_y = \frac{dr}{r} = \frac{dt}{t-t_0} - id\theta,$$

where t and t_0 are the affixes of the points y and z, the equation (I_1) can be written in the form

$$\psi(t_0) + K^* \int_L \frac{\psi(t)dt}{t-t_0} + F\psi = g(t_0),$$

where F is a Fredholm operator. It follows from (2.11.34) that the system (I_1) is a system of singular integral equations for which Fredholm's basic theorems are valid. It can be proved that Fredholm's basic theorems are also applicable to the system (I_2).

δ) Existence theorems. Let us consider the problems (I_1) and (E_2). The homogeneous equations (I_1^o) and (E_2^o) have only trivial solutions. We assume the opposite and suppose that ψ_o is a solution of equation (E_2^o), not equal to zero. If we multiply the equation (E_2^o), by ds_z and integrate on L we obtain

$$\int_L \psi_o(z)ds_z = 0.$$

Thus, the single layer potential

$$V_o(x;\psi_o) = \int_L \Gamma(x,y)\psi_o(y)ds_y$$

satisfies the conditions (2.11.43) and

$$A(\frac{\partial}{\partial x})V_o = 0 \text{ on } \Sigma_e, \; [H(\frac{\partial}{\partial z}, n_z)V_o]_e = 0 \text{ on } L.$$

According to Theorem 2.11.4 we obtain $V_o = 0$ on Σ_e. By Theorem 2.11.1,

$$[V_0(x;\psi_0)]_e = [V_0(x;\psi_0)]_i = 0, \quad x \in L.$$

Thus, we have

$$A(\frac{\partial}{\partial x})V_0 = 0 \text{ on } \Sigma, \quad V_0 = 0 \text{ on } L.$$

It follows from Theorem 2.11.4 that $V_0 = 0$ on Σ. Thus, by Theorem 2.11.3,

$$\psi(z) = [H(\frac{\partial}{\partial z}, n_z)V_0]_i - [H(\frac{\partial}{\partial z}, n_z)V_0]_e = 0 \text{ on } L.$$

We arrived at a contradiction which shows that the statement concerning equation (E_2^0) is valid.

Since the equations (I_1^0) and (E_2^0) are an associate set of singular equations, according to the second Fredholm theorem, (I_1^0) has also no non-trivial solutions. It follows from (E_2) that

$$\int_L \psi(z)ds_z = -\int_L g(z)ds_z.$$

Thus we have

Theorem 2.11.5. The problem (I_1) has solution for any vector field g of class H. This solution is unique and can be expressed by a double layer potential.

Theorem 2.11.6. The problem (E_2) has solution for any vector field g of class H if and only if

$$\int_L g(z)ds_z = 0.$$

This solution is unique and can be expressed by a single layer potential.

Let us consider the equations (I_2^0) and (E_1^0). Clearly, the vector u^0 defined by (2.11.22) satisfies the relations

$$A(\frac{\partial}{\partial x})u^0 = 0, \quad H(\frac{\partial}{\partial x}, n_x)u^0 = 0.$$

It follows form (2.11.36) that

$$u^o = -\int_L \Lambda(x,y)u^o ds_y.$$

Passing to the limit as the point x approaches the boundary point $z \in L$ from within, according to Theorem 2.11.2 we have

$$\frac{1}{2} u^o + \int_L \Lambda(z,y)u^o ds_y = 0.$$

Hence u^o satisfies the equation (E_1^o). In the expression for u^o we shall take successively one of the constants different from zero and the other two equal to zero. Thus, we are led to the vector set $u^{(1)} = (1,0,0)$, $u^{(2)} = (0,1,0)$, $u^{(3)} = (0,0,1)$, forming linearly independent solutions of equation (E_1^o). According to Fredholm's second theorem, the associate equation (I_2^o) has at least three linearly independent solutions $\xi^{(j)}$ (J = 1, 2,3). As in [237], we can show that $\xi^{(j)}$ form a complete system of linearly independent solutions of the equation (I_2^o). From this we deduce the completeness of the associate system $u^{(j)}$ (j = 1,2,3). In agreement with Fredholm's third theorem, the necessary and sufficient conditions to solve the equation (I_2) are

$$\int_L u^{(j)}g\ ds = 0 \quad (j = 1,2,3).$$

If we take $g = (\tilde{s}_1,\tilde{s}_2,\tilde{s}_3)$, then the above conditions can be written in the form

$$\int_L \tilde{s}_i\ ds = 0. \tag{2.11.44}$$

Thus, we have

<u>Theorem 2.11.7</u>. The problem (I_2) has solution for any vector field of class H, if and only if the conditions (2.11.44) are fulfilled. The solution can be expressed as a single layer potential.

As in the classical theory, we can prove

<u>Theorem 2.11.8</u>. The problem (E_1) has solution for any vector field g of class H and the solution can be expressed as a linear combination of a double layer potential and some single layer potentials.

3 Elastodynamics

3.1. Elastic processes

We now assume that the motion of the body is referred to the configuration B_0, occupied by the body at time $t_0 = 0$. We say that the vector field $\underline{u}$ is an admissible displacement field on $\bar{B}_0 \times [0,t_1)$ provided: (i) $\underline{u}$ is of class C^2 on $B_0 \times (0,t_1)$; (ii) $\underline{u}$, $\dot{\underline{u}}, \ddot{\underline{u}}, \underline{u}_{,A}$ and $\dot{\underline{u}}_{,A}$ are continuous on $\bar{B}_0 \times [0,t_1)$.

The fundamental system of field equations describing the incremental motion consists of the stress-displacement relations

$$S_{Ai} = D_{iAjB}u_{j,B}, \tag{3.1.1}$$

and the equations of motion

$$S_{Ai,A} + \rho_0 F_i = \rho_0 \ddot{u}_i, \tag{3.1.2}$$

on $B_0 \times (0,t_1)$. We assume that

(i) D_{iAjB} are continuously differentiable on $\bar{B}_0 \times [0,t_1)$ and satisfy the symmetry relations (1.2.38);

(ii) ρ_0 is continuous and strictly positive on $\bar{B}_0$;

(iii) $\underline{F}$ is continuous on $\bar{B}_0 \times [0,t_1)$.

An admissible stress field on $\bar{B}_0 \times [0,t_1)$ is a tensor field S_{Ai} with the properties: (i) S_{Ai} are of class $C^{1,0}$ on $B_0 \times (0,t_1)$; (ii) S_{Ai} and $S_{Ai,A}$ are continuous on $\bar{B}_0 \times [0,t_1)$. By an admissible process we mean an ordered array of functions $p = (u_i, S_{Aj})$ with the properties: (α) $\underline{u}$ is an admissible displacement field on $\bar{B}_0 \times [0,t_1)$; (β) S_{Ai} is an admissible stress field. Here, the domain of definition of an admissible process is understood to be $\bar{B}_0 \times [0,t_1)$.

We say that $p = (u_i, S_{Aj})$ is an elastic process corresponding to the body force $\underline{F}$ if p is an admissible process that satisfies the stress-displacement relations (3.1.1) and the equations of motion (3.1.2). The surface traction $\underline{P}$ is defined at every regular point of $\partial B_0 \times [0,t_1)$ by $P_i = S_{Ai}N_A$. When we omit mention of the domain of definition of an elastic process, it will always be understood to be $\bar{B}_0 \times [0,t_1)$.

Given an elastic process, we define the functions E and U on $[0,t_1)$ by

$$E = \frac{1}{2}\int_{B_0} (\rho_0\dot{u}_i\dot{u}_i + D_{iAjB}u_{i,A}u_{j,B})dV, \tag{3.1.3}$$

$$U = E - \frac{1}{2}\int_0^t \int_{B_0} \dot{D}_{iAjB}u_{i,A}u_{j,B}\, dV\, ds. \tag{3.1.4}$$

Theorem 3.1.1. Let (u_i,S_{Aj}) be an elastic process corresponding to the body force $\underline{F}$. Then

$$\int_{\partial B_0} P_i\dot{u}_i dA + \int_{B_0} \rho_0 F_i\dot{u}_i\, dV = \dot{U}. \tag{3.1.5}$$

Proof. By (3.1.2),

$$S_{Ai}\dot{u}_{i,A} = (S_{Ai}\dot{u}_i)_{,A} + \rho_0 F_i\dot{u}_i - \rho_0\dot{u}_i\ddot{u}_i. \tag{3.1.6}$$

It follows from (1.2.38), (3.1.1) and (3.1.6) that

$$\frac{1}{2}\frac{\partial}{\partial t}(\rho_0\dot{u}_i\dot{u}_i + D_{iAjB}u_{i,A}u_{j,B}) = (S_{Ai}\dot{u}_i)_{,A} + \rho_0 F_i\dot{u}_i + \\ + \frac{1}{2}\dot{D}_{iAjB}u_{i,A}u_{j,B}. \tag{3.1.7}$$

By the divergence theorem and (3.1.7),

$$\dot{E} - \frac{1}{2}\int_{B_0} \dot{D}_{iAjB}u_{i,A}u_{j,B}\, dV = \int_{\partial B_0} P_i\dot{u}_i\, dA + \int_{B_0} \rho_0 F_i\dot{u}_i\, dV. \tag{3.1.8}$$

The relations (3.1.4) and (3.1.8) imply the desired result. □

The following proposition is an immediate consequence of Theorem 3.1.1.

Theorem 3.1.2. Let (u_i, S_{Aj}) be an elastic process corresponding to the body force $\underline{F}$, and suppose that

$$F_i = 0 \text{ on } \bar{B}_0 \times [0,t_1), \quad P_i\dot{u}_i = 0 \text{ on } \partial B_0 \times [0,t_1).$$

Then U is conserved, i.e.

$$U(t) = U(0), \quad 0 \leq t \leq t_1.$$

Note that the relation (3.1.5) may be written in the form

$$\int_0^t \int_{\partial B_0} P_i\dot{u}_i \, dA \, ds + \int_0^t \int_{B_0} \rho_0 F_i \dot{u}_i \, dV \, ds = U(t) - U(0). \tag{3.1.9}$$

If the primary state is a configuration of equilibrium, then E = U.

Let f and g be scalar fields on $B_0 \times [0,t_1)$ that are continuous in time. By the convolution of f and g we mean the function $f * g$ defined on $B_0 \times [0,t_1)$ by

$$f * g(\underline{X},t) = \int_0^t f(\underline{X},t-\tau)g(\underline{X},\tau)d\tau. \tag{3.1.10}$$

Henceforth (unless otherwise specified) i will denote the function on $[0,t_1)$ defined by

$$i(t) = t. \tag{3.1.11}$$

Let $\hat{\underline{a}}$ and $\hat{\underline{b}}$ be vector fields on $\bar{B}_0$, and suppose that $\underline{F}$ is a continuous vector field on $B_0 \times [0,t_1)$. We denote by $\underline{g}$ the vector field on $\bar{B}_0 \times [0,t_1)$ defined by

$$\underline{g} = \rho_0 i * \underline{F} + \rho_0(\hat{\underline{a}} + t\hat{\underline{b}}). \tag{3.1.12}$$

Theorem 3.1.3. Let $p = (u_i, S_{Ai})$ be an admissible process. Then p satisfies the equations of motion (3.1.2) and the initial conditions $\underline{u}(.,0) = \hat{\underline{a}}$, $\dot{\underline{u}}(.,0) = \hat{\underline{b}}$ if and only if

$$i * S_{Aj,A} + g_j = \rho_0 u_j. \tag{3.1.13}$$

Proof. Assume that (3.1.2) and the initial conditions hold. Taking the convolution of the equations of motion with i, we conclude, with the aid of (3.1.12), that the equation (3.1.13) is met. The converse is easily verified by reversing the foregoing argument. □

This theorem is due to Ignaczak [201] and Gurtin [152]. It will be useful in establishing reciprocity relations and variational theorems.

3.2 Boundary-initial-value problems

The fundamental system of field equations for linear elastodynamics is given by (3.1.1) and (3.1.2). To this system we adjoin the *initial conditions*

$$\underline{u}(.,0) = \hat{\underline{a}},\ \dot{\underline{u}}(.,0) = \hat{\underline{b}} \text{ on } \bar{B}_0, \tag{3.2.1}$$

the *displacement condition*

$$\underline{u} = \hat{\underline{u}} \text{ on } \bar{\Sigma}_1 \times [0,t_1), \tag{3.2.2}$$

and the *traction condition*

$$\underline{P} = \hat{\underline{P}} \text{ on } \bar{\Sigma}_2 \times [0,t_1). \tag{3.2.3}$$

Here $\hat{\underline{a}}$ and $\hat{\underline{b}}$ are prescribed initial displacement and initial velocity, while $\hat{\underline{u}}$ and $\hat{\underline{P}}$ are given surface displacement and surface traction. We assume that

(i) $\hat{\underline{a}}$ and $\hat{\underline{b}}$ are continuous on $\bar{B}_0$;

(ii) $\hat{\underline{u}}$ is continuous on $\bar{\Sigma}_1 \times [0,t_1)$;

(iii) $\hat{\underline{P}}$ is continuous in time and piecewise regular on $\Sigma_2 \times [0,t_1)$;

(iv) $\underline{F}$ is continuous on $\bar{B}_0 \times [0,t_1)$.

The mixed problem of elastodynamics consists in finding an elastic process that corresponds to the body force $\underline{F}$ and satisfies the initial conditions (3.2.1) and the boundary conditions (3.2.2), (3.2.3). When $\Sigma_1 = \partial B_0$ the associated problem is called the displacement problem. If $\Sigma_2 = \partial B_0$ we refer to the resulting problem as the traction problem.

By an *external data system* on $\bar{B}_0 \times [0,t_1)$ for the mixed problem we mean an ordered array $L = (\underline{F},\hat{\underline{u}},\hat{\underline{P}},\hat{\underline{a}},\hat{\underline{b}})$ with properties (i)-(iv).

Substitution of (3.1.1) into the equations (3.1.2) and (3.2.3) yields the equations

$$(D_{iAjB}u_{j,B})_{,A} + {}_{o}F_i = \rho_0 \ddot{u}_i \text{ on } B_o \times (0,t_1), \tag{3.2.4}$$

and the boundary conditions

$$u_i = \hat{u}_i \text{ on } \bar{\Sigma}_1 \times [0,t_1),\ D_{iAjB}u_{j,B}N_A = \hat{P}_i \text{ on } \Sigma_2 \times [0,t_1) \tag{3.2.5}$$

By a *solution* of the mixed problem we mean an admissible displacement field $\underline{u}$ that satisfies the equations (3.2.4), the initial conditions (3.2.1) and the boundary conditions (3.2.5).

When the primary state is a configuration of equilibrium then we refer all quantities to this configuration. In this case the fundamental system of field equations consists of the stress-displacement relations

$$s_{ji} = d_{ijrs}u_{r,s}, \tag{3.2.6}$$

and the equations of motion

$$s_{ji,j} + \rho F_i = \rho \ddot{u}_i, \tag{3.2.7}$$

on $B \times (0,t_1)$. We assume that d_{ijrs} are continuously differentiable on $\bar{B}$ and satisfy the symmetry relations (1.2.45), and that the density ρ is continuous and strictly positive on $\bar{B}$. The surface traction $\underline{s}$ is defined at every regular point of $\partial B \times [0,t_1)$ by $s_i = s_{ji}n_j$. To the system of field equations just cited we adjoin the initial conditions

$$\underline{u}(.,0) = \tilde{\underline{a}},\ \dot{\underline{u}}(.,0) = \tilde{\underline{b}} \text{ on } \bar{B}, \tag{3.2.8}$$

and the boundary conditions

$$\underline{u} = \tilde{\underline{u}} \text{ on } \bar{S}_1 \times [0,t_1),\ \underline{s} = \tilde{\underline{p}} \text{ on } S_2 \times [0,t_1). \tag{3.2.9}$$

As before, we can introduce the external data system on $\bar{B} \times [0,t_1)$,

$\tilde{L} = (\underline{F}, \tilde{\underline{u}}, \tilde{\underline{p}}, \tilde{\underline{a}}, \tilde{\underline{b}})$. Substitution of (3.2.6) into (3.2.7) and (3.2.9) yields the equations

$$(d_{ijrs}u_{r,s})_{,j} + \rho F_i = \rho \ddot{u}_i \text{ on } B \times (0,t_1), \tag{3.2.10}$$

and the boundary conditions

$$u_i = \tilde{u}_i \text{ on } \bar{S}_1 \times [0,t_1), \quad d_{ijrs}u_{r,s}n_j = \tilde{p}_i \text{ on } S_2 \times [0,t_1). \tag{3.2.11}$$

In this case, by a solution of the mixed problem we mean an admissible displacement field on $\bar{B} \times [0,t_1)$ that satisfies the equations (3.2.10), the initial conditions (3.2.8) and the boundary conditions (3.2.11).

It follows from (1.2.51) and (3.1.3) that the function E may be written in the form

$$E = \frac{1}{2}\int_B (\rho \dot{u}_i \dot{u}_i + d_{ijrs}u_{i,j}u_{r,s})dv. \tag{3.2.12}$$

When the primary state is time-independent, then E = U. Theorem 3.1.2 has the following consequence

Theorem 3.2.1. Let $\underline{u}$ be a solution of the mixed problem on $\bar{B} \times [0,t_1)$ and suppose that

$$\underline{F} = \underline{0} \text{ on } B \times [0,t_1), \ \underline{s}.\dot{\underline{u}} = 0 \text{ on } \partial B \times [0,t_1).$$

Then E is conserved, i.e.

$$E(t) = E(0), \quad 0 \leq t < t_1. \tag{3.2.13}$$

Let $\underline{h}$ be the vector field on $\bar{B} \times [0,t_1)$ defined by

$$\underline{h} = \rho(i * \underline{F} + \tilde{\underline{a}} + t\tilde{\underline{b}}). \tag{3.2.14}$$

Theorem 3.2.2. Let $p = (u_i, s_{ji})$ be an admissible process on $\bar{B} \times [0,t_1)$. Then p satisfies the equations of motion (3.2.7) and the initial conditions (3.2.8) if and only if

$$i * s_{kj,k} + h_j = \rho u_j. \tag{3.2.15}$$

The proof of this theorem is strictly analogous to that of Theorem 3.1.3 and can safely be omitted.

The next proposition is an immediate consequence of Theorem 3.2.2.

Theorem 3.2.3. Let $\underline{u}$ be an admissible displacement field on $\bar{B} \times [0,t_1)$. Then $\underline{u}$ is a solution of the mixed problem if and only if

$$i * (d_{jkrs}u_{r,s})_{,k} + \rho h_j = \rho u_j \text{ on } B \times [0,t_1),$$

$$u_i = \tilde{u}_i \text{ on } \bar{S}_1 \times [0,t_1), \quad d_{ijrs}u_{r,s}n_j = \tilde{p}_i \text{ on } S_2 \times [0,t_1).$$

Theorem 3.2.3 furnishes an alternative characterization of the solution to the mixed problem in which the initial conditions are incorporated into the field equations.

3.3 Uniqueness and continuous dependence results

In a series of papers [222-225], Knops and Payne have studied questions of continuous dependence and uniqueness for solutions of various classes of initial boundary value problems of elastodynamics, in the case when the primary state is a configuration of equilibrium. These investigations have been established by using logarithmic convexity arguments, with no definiteness assumption on the elasticities. In [226], the convexity method has been used by Knops and Payne to derive uniqueness and continuous dependence results in thermoelastodynamics. Galdi and Rionero [128], [129] have obtained uniqueness and continuous dependence results in elastodynamics for exterior domains by using the weight function method. Knops and Payne [230], by coupling convexity and weight function methods have established uniqueness and continuous dependence results in nonlinear elastodynamics. Stability of an elastic body with time-dependent elasticities has been studied by Knops and Wilkes [229]. Chiriţa [79] has obtained uniqueness and continuous dependence results in elastodynamics for time-dependent primary states by using the logarithmic convexity and weight function methods.

In this section we establish some uniqueness and continuous dependence results of elastodynamics. An existence result can be derived by using the theory of semigroups of linear operators (see Section 5.5).

α) Time-independent primary states. We assume that the primary state is a configuration of equilibrium. Moreover, we suppose that B is a bounded regular region of three-dimensional Euclidean space. We present a uniqueness theorem which was established independently by Brun [40], [41] and Knops and Payne [222], using different methods.

Theorem 3.3.1. Let the primary state be time-independent. Then the mixed problem of elastodynamics has at most one solution.

Proof. We first present the proof given by Brun [40], [41]. Let us consider $r_1, r_2 \in (0, t_1)$. On the basis of the symmetry relations (1.2.45) we have

$$s_{ji}(r_1)u_{i,j}(r_2) = s_{ji}(r_2)u_{i,j}(r_1), \tag{3.3.1}$$

where, for convenience, we have suppressed the argument $\underline{x}$. If we define

$$D_{\alpha\beta} = \int_B s_{ji}(r_\alpha)u_{i,j}(r_\beta)dv, \tag{3.3.2}$$

we are led, with the aid of (3.3.1), to the relation

$$D_{12} = D_{21}. \tag{3.3.3}$$

By (3.2.8),

$$\begin{aligned} s_{ji}(r_\alpha)u_{i,j}(r_\beta) = (s_{ji}(r_\alpha)u_i(r_\beta))_{,j} - \rho\ddot{u}_i(r_\alpha)u_i(r_\beta) + \\ + \rho F_i(r_\alpha)u_i(r_\beta). \end{aligned} \tag{3.3.4}$$

The divergence theorem and (3.3.4) imply

$$\begin{aligned} D_{\alpha\beta} = \int_{\partial B} s_{ji}(r_\alpha)n_j u_i(r_\beta)da + \int_B \rho(F_i(r_\alpha)u_i(r_\beta) - \\ - \ddot{u}_i(r_\alpha)u_i(r_\beta))dv. \end{aligned} \tag{3.3.5}$$

By (2.1.3), (3.3.3) and (3.3.5),

$$\int_{\partial B} [s_i(r_1)u_i(r_2) - s_i(r_2)u_i(r_1)]da = \int_B \rho[F_i(r_1)u_i(r_2) - F_i(r_2)u_i(r_1) - \ddot{u}_i(r_1)u_i(r_2) + \ddot{u}_i(r_2)u_i(r_1)]dv. \tag{3.3.6}$$

Let $\underline{u}$ be the difference between two solutions of the mixed problem. Then $\underline{u}$ corresponds to vanishing body forces and satisfies

$$\underline{u}(.,0) = 0,\ \dot{\underline{u}}(.,0) = 0 \text{ on } \bar{B},\ \underline{u} = \underline{0} \text{ on } \bar{S}_1 \times [0,t_1),$$
$$\underline{s} = \underline{0} \text{ on } S_2 \times [0,t_1). \tag{3.3.7}$$

It follows from (3.3.6) and (3.3.7) that

$$\int_B \rho[\ddot{u}_i(r_1)u_i(r_2) - \ddot{u}_i(r_2)u_i(r_1)]dv = 0.$$

Next, if we take $r_1 = \eta$, $r_2 = 2\tau-\eta$ where $2\tau > \eta$ and integrate from 0 to τ, we arrive at

$$\int_0^\tau \int_B \rho[u_i(2\tau-\eta)\ddot{u}_i(\eta) - u_i(\eta)\ddot{u}_i(2\tau-\eta)]dv\, d\eta = 0.$$

Integrating by parts with respect to time and using the relations (3.3.7), we obtain

$$\int_B \rho u_i\dot{u}_i dv = 0.$$

By a further integration and (3.3.7) we conclude that

$$\int_B \rho u_i u_i dv = 0,$$

from which uniqueness easily follows.

We now present the proof due to Knops and Payne [222]. The method depends upon convexity arguments. The behaviour of the body is studied on a finite time interval. Let $\underline{u}$ be a solution of the mixed problem corresponding to vanishing body forces and satisfying the conditions (3.3.7).

It follows from (3.2.13) and (3.3.7) that $E(0) = 0$. Moreover, by Theorem 3.2.1 we have

$$E = 0. \tag{3.3.8}$$

Following [222] we define the function F on $[0,t_1]$ by

$$F = \int_B \rho u_i u_i \, dv. \tag{3.3.9}$$

Clearly, the uniqueness is implied by $F = 0$ on $[0,t_1]$. Hence, we suppose that there is an interval (t_2,t_3), $0 \leq t_2$, $t_3 \leq t_1$, such that $F(t) > 0$, $t \in (t_2,t_3)$. Let us prove that

$$F\ddot{F} - \dot{F}^2 > 0 \quad \text{on} \quad (t_2,t_3). \tag{3.3.10}$$

Clearly,

$$\dot{F} = 2 \int_B \rho u_i \dot{u}_i \, dv,$$

$$\ddot{F} = 2 \int_B \rho(\dot{u}_i\dot{u}_i + u_i\ddot{u}_i)dv. \tag{3.3.11}$$

In view of the equations of motion,

$$\rho u_i\ddot{u}_i = u_i s_{ji,j} = (u_i s_{ji})_{,j} - s_{ji}u_{i,j}. \tag{3.3.12}$$

By (3.2.7), (3.3.12), the divergence theorem and (3.3.7) we obtain

$$\int_B \rho u_i\ddot{u}_i dv = - \int_B d_{ijrs}u_{i,j}u_{r,s}dv. \tag{3.3.13}$$

It follows from (3.3.8), (3.3.11) and (3.3.13) that

$$\ddot{F} = 4 \int_B \rho \dot{u}_i\dot{u}_i dv.$$

Thus, we arrive at

$$F\ddot{F} - \dot{F}^2 = 4 \left\{\int_B \rho u_i u_i dv \int_B \rho\dot{u}_i\dot{u}_i dv - \left(\int_B \rho u_i\dot{u}_i dv\right)^2\right\}.$$

It then follows by Schwarz's inequality that (3.3.10) holds. Without loss we may take $F(t_2) = 0$. Since

$$F\ddot{F} - \dot{F}^2 = F^2 \frac{d^2}{dt^2} (\ln F), \ t_2 < t < t_3,$$

we conclude that ln F is a convex function on (t_2, t_3). Together with the continuity of F this implies

$$F(t) \leq F(t_2)^{w_1} F(t_3)^{w_2}, \quad t \in (t_2, t_3), \tag{3.3.14}$$

where

$$w_1 = (t_3 - t)/(t_3 - t_2), \quad w_2 = (t - t_2)/(t_2 - t_3).$$

Since $F(t_2) = 0$, the relation (3.3.14) implies that $F = 0$ on $[0, t_3)$, and by continuity $F(t_3) = 0$. We conclude that $F = 0$ on $[0, t_1]$. Then, since $\underline{u} \in C^2(B \times [0, t_1])$, we obtain $\underline{u} = \underline{0}$ on $\bar{B} \times [0, t_1]$. □

β) <u>Time-dependent primary states</u>. The results we give here are due to Chiriţa [79]. As before, the behaviour of the body is investigated on a finite time interval.

We suppose that ρ_0 is continuous on B_0, and that D_{iAjB} are continuously differentiable on $\bar{B}_o \times [0, t_1]$. For the remainder of this section we assume that there exist the constants λ_i such that

$$\dot{D}_{iAjB} u_{i,A} u_{j,B} \leq \lambda_1 \rho_0 u_i u_i + \lambda_2 \rho_0 \dot{u}_i \dot{u}_i + \lambda_3 D_{iAjB} u_{i,A} u_{j,B} \tag{3.3.15}$$

on $B_o \times [0, t_1]$, for all displacement fields $\underline{u}$. Clearly, this condition is identically satisfied when the primary state is a configuration of equilibrium. The condition (3.3.15) is satisfied when $\dot{D}_{iAjB}$ is negative definite. Moreover, it is also satisfied if D_{iAjB} is positive or negative definite.

Let $\underline{u}^{(1)}$ and $\underline{u}^{(2)}$ be solutions of the mixed problem corresponding to the external data systems $L^{(1)} = (\underline{F}^{(1)}, \hat{\underline{u}}, \underline{P}, \hat{\underline{a}}^{(1)}, \hat{\underline{b}}^{(1)})$ and $L^{(2)} = (\underline{F}^{(2)}, \hat{\underline{u}}, \underline{P}, \hat{\underline{a}}^{(2)}, \hat{\underline{b}}^{(2)})$, respectively. If we define $\underline{u} = \underline{u}^{(1)} - \underline{u}^{(2)}$, then $\underline{u}$ is a solution of the initial-boundary-value problem corresponding to the external data system $L = (\underline{F}, \underline{0}, \underline{0}, \hat{\underline{a}}, \hat{\underline{b}})$ where $\underline{F} = \underline{F}^{(1)} - \underline{F}^{(2)}$, $\hat{\underline{a}} = \underline{a}^{(1)} - \underline{a}^{(2)}$, $\hat{\underline{b}} = \hat{\underline{b}}^{(1)} - \hat{\underline{b}}^{(2)}$.

This problem is denoted by (A). Let D_A be the class of the solutions to the problem (A).

Let i be the function on $[0,\infty)$ defined by $i(t) = t$. Then for any function f continuous in time,

$$i * f = \int_0^t (t - \eta)f(\eta)d\eta.$$

Clearly,

$$\int_0^t \int_0^\eta f(\xi)d\xi \; d\eta = \int_0^t (t - \eta)f(\eta)d\eta = i * f.$$

For X^o a point of E^3 and α a positive constant, let g denote the function defined by

$$g(\underline{X}) = \exp(-\alpha|\underline{X} - \underline{X}^o|), \quad \underline{X} \in E^3.$$

Let $f \in C^2(E^3)$ be a bounded positive function with bounded derivatives such that $f(\underline{z}) = 1$ for $|\underline{z}| \leqq 1$, and $f(\underline{z}) = 0$ for $|\underline{z}| \geqq 2$, and let

$$f^o(\underline{X}) = f[(\underline{X} - \underline{X}^o)/R], \; R > 0.$$

We define the function h by

$$h = gf^o. \tag{3.3.16}$$

Let y, K and W be the functions on $[0,t_1)$ defined by

$$y = \int_0^t \int_{B_0} \rho_0 h(\dot{u}_i\dot{u}_i - u_i\ddot{u}_i)dVd\eta, \; K = \frac{1}{2}\int_{B_0} \rho_0 h\dot{u}_i\dot{u}_i dV,$$

$$W = \frac{1}{2}\int_B h \; S_{Ai}(\underline{u})u_{i,A} \; dV, \tag{3.3.17}$$

where $\underline{u} \in D_A$ and $S_{Ai}(\underline{u}) = D_{iAjK}u_{j,K}$.

Lemma 3.3.1. Let $\underline{u}$ be a solution of the problem (A). Then

$$\dot{y} = 2K(0) + 2W(0) - Q_o + H_o + 1 * (V - 2H_1 + 2Q_1), \tag{3.3.18}$$

$$\dot{y} = 2K + 2W - Q_o + H_o, \tag{3.3.19}$$

$$2i * (K-W) = 1 * G - tG(0) + i * (H_o - Q_o), \tag{3.3.20}$$

where

$$V = \int_{B_o} h\dot{D}_{iAjK}u_{i,A}u_{j,K}dV, \qquad Q_o = \int_{B_o} \rho_o hF_i u_i dV,$$

$$Q_1 = \int_{B_o} \rho_o hF_i\dot{u}_i dV, \qquad H_o = \int_{B_o} D_{iAjK}u_i h_{,A}u_{j,K}dV, \tag{3.3.21}$$

$$H_1 = \int_{B_o} D_{iAjK}\dot{u}_i h_{,A}u_{j,K}\, dV, \quad G = \int_{B_o} \rho_o hu_i\dot{u}_i dV.$$

<u>Proof</u>. By (3.1.2),

$$hS_{Ai}\dot{u}_{i,A} = (hS_{Ai}\dot{u}_i)_{,A} + \rho_o hF_i\dot{u}_i - h_{,A}S_{Ai}\dot{u}_i - \rho_o h\dot{u}_i\ddot{u}_i. \tag{3.3.22}$$

It follows from (1.2.38), (3.1.1) and (3.3.22) that

$$\frac{1}{2}\frac{\partial}{\partial t}[h(\rho_o\dot{u}_i\dot{u}_i + D_{iAjK}u_{i,A}u_{j,K})] = (hS_{Ai}\dot{u}_i)_{,A} + \rho_o hF_i\dot{u}_i - \\ - D_{iAjK}\dot{u}_i h_{,A}u_{j,K} + \frac{1}{2}h\dot{D}_{iAjK}u_{i,A}u_{j,K}. \tag{3.3.23}$$

Similarly,

$$hS_{Ai}u_{i,A} = (hS_{Ai}u_i)_{,A} + \rho_o hF_i u_i - h_{,A}S_{Ai}u_i - \rho_o hu_i\ddot{u}_i,$$

so that

$$\rho_o hu_i\ddot{u}_i = (hS_{Ai}u_i)_{,A} + \rho_o hF_i u_i - D_{iAjK}u_i h_{,A}u_{j,K} - \\ - hD_{iAjK}u_{i,A}u_{j,K}. \tag{3.3.24}$$

By the divergence theorem, the relations (3.3.23) and (3.3.24) imply

$$\dot{K} + \dot{W} = Q_1 - H_1 + \frac{1}{2} V, \tag{3.3.25}$$

and

$$\dot{y} = 2K - Q_o + H_o + 2W, \tag{3.3.26}$$

respectively. The relation (3.3.25) is equivalent to

$$K + W = K(0) + W(0) + 1 * (Q_1 - H_1 + \frac{1}{2} V). \tag{3.3.27}$$

Clearly, (3.3.26) and (3.3.27) imply (3.3.18). Since $\dot{\overline{u_i\dot{u}_i}} = \dot{u}_i\dot{u}_i + u_i\ddot{u}_i$, we find that

$$G(t) - G(0) = 2 * K + \int_0^t \int_{B_o} \rho_o h u_i \ddot{u}_i \, dV \, ds.$$

The above relation may be written in the form

$$G(t) - G(0) = 4 * K - y. \tag{3.3.28}$$

The relations (3.3.26) and (3.3.28) imply (3.3.20). □

Let F be the function on $[0,t_1)$ defined by

$$F = 2 * L + 2(t_1 - t)L(0) + M^2, \tag{3.3.29}$$

where

$$L = \frac{1}{2}\int_{B_o} \rho_o g u_i u_i dV,$$

$$M^2 = 2L(0) + \int_{B_o} \rho_o g \hat{b}_i \hat{b}_i dV + 2\int_{B_o} g|w(\underline{X},0)|dV + \int_0^{t_1}\int_{B_o} \rho_o g[F_i F_i + \alpha^2 \rho_o^{-2} S_{Ai}(\underline{u})S_{Ai}(\underline{u})]dV \, d\tau, \tag{3.3.30}$$

$$2w = S_{Ai}(\underline{u})u_{i,A},$$

and $\underline{u} \in D_A$. We introduce the notation $q = (\rho_0 g)^{1/2}$.

Theorem 3.3.2. Let $\underline{u}$ be a solution of the problem (A), and assume that

$$qF_i, qu_i, q\dot{u}_i, \rho_0^{-1} qS_{Ai}(\underline{u}) \in L_2(B_0 \times [0,t_1)), \quad q\hat{a}_i, \; q\hat{b}_i \in L_2(B_0),$$

$$g|w(.,0)| \in L_1(B_0),$$

for any $\alpha \in (0,1]$. Then for finite time t_1 and for all $\underset{\sim}{X}^o \in E^3$ it is possible to determine the positive constants m_1 and m_2 such that

$$F(t) \leq [F(0)]^{\delta}[F(t_1)\exp(m_1 t_1/m_2)]^{1-\delta} \exp(-m_1 t/m_2), \tag{3.3.31}$$

where

$$\delta = [\exp(-m_2 t) - \exp(-m_2 t_1)]/[1 - \exp(-m_2 t_1)]. \tag{3.3.32}$$

Proof. We define the function P on $[0,t_1)$ by

$$P = 2 * D + 2(t_1 - t)D(0) + c^2, \tag{3.3.33}$$

where

$$\begin{aligned} D &= \frac{1}{2}\int_{B_0} \rho_0 h u_i u_i dV, \quad c^2 = 2K(0) + 2D(0) + 2\int_{B_0} h|w(\underset{\sim}{X},0)|dV + \\ &+ \int_0^{t_1}\int_{B_0} \rho_0 h[F_i F_i + \alpha^2 \rho_0^{-2} S_{Ai}(\underline{u})S_{Ai}(\underline{u})]dV\, d\tau + \\ &+ \int_0^{t_1}\int_{B_0} g[|S_{Ai}(\underline{u})f^o{}_{,A} u_i| + |S_{Ai}(\underline{u})f^o{}_{,A} \dot{u}_i|]dV\, d\tau. \end{aligned} \tag{3.3.34}$$

Clearly, by (3.3.17), (3.3.19), (3.3.20) and (3.3.34)

$$\begin{aligned} \dot{P} &= 2D - 2D(0) = 2 * G, \\ \ddot{P} &= 2G = 8 * K - 2y + 2G(0). \end{aligned} \tag{3.3.35}$$

By using Schwarz's and Cauchy's inequalities we obtain

$$-2G(0) \leqq 2D(0) + 2K(0) \leqq C^2. \qquad (3.3.36)$$

The relations (3.3.18) and (3.3.19) can be written in the form

$$i * V = y - 2t[K(0) + W(0)] + 1 * (Q_o - H_o) + \\ + 2i * (H_1 - Q_1) \qquad (3.3.37)$$

and

$$2 * (K + W) = y + 1 * (Q_o - H_o), \qquad (3.3.38)$$

respectively. It follows from (3.3.20) and (3.3.38) that

$$4i * K - 1 * y = 1 * G - tG(0). \qquad (3.3.39)$$

By (3.3.15), (3.3.37) and (3.3.38),

$$y \leqq \lambda_3(1 * y) + i * (2 + |\lambda_1|D+2|\lambda_2-\lambda_3|K + \lambda_3 Q_o + 2Q_1 - \\ -2H_1 - \lambda_3 H_o) + 1 * (H_o-Q_o) + 2t(K(0) + W(0)). \qquad (3.3.40)$$

If we multiply the relation (3.3.39) by the arbitrary positive constant γ and add the result to (3.3.40), then we arrive at

$$y \leqq (\gamma + \lambda_3)(1*y) + i * [2|\lambda_1|D + 2(|\lambda_2-\lambda_3|-2\gamma)K + \lambda_3 Q_o + \\ + 2Q_1 - 2H_1 - \lambda_3 H_o) + 1 * (H_o-Q_o) + 2t(K(0) + W(0)) + \\ + \gamma[1 * G - tG(0)]. \qquad (3.3.41)$$

Using Schwarz's and Cauchy's inequalities and the estimate

$$i * \phi \leqq t_1(1 * |\phi|),$$

we find that

$$
\begin{aligned}
y \leqq (\gamma + \lambda_3)(1 * y) &+ 2(|\lambda_2 - \lambda_3| - 2\gamma + 2)i * K + \\
&+ \gamma[(2*D)(2*K)]^{1/2} + [(|\lambda_1| + |\lambda_3|)t_1 + 1](2*D) + \\
&+ (\gamma + 2)t_1[D(0) + K(0) + \int_{B_o} h|w(\underline{X},0)|dV] + \\
&+ \frac{1}{2}[(|\lambda_3|+2)t_1+1] \int_0^{t_1} \int_{B_o} \rho_o h[F_i F_i + \alpha^2 \rho_o^{-2} S_{Ai}(\underline{u}) S_{Ai}(\underline{u})]dVd\tau + \\
&+ \int_0^{t_1} \int_{B_o} g[2t_1|S_{Ai}(\underline{u})f^o{}_{,A}\dot{u}_i| + (1+|\lambda_3|t_1)|S_{Ai}(\underline{u})f^o{}_{,A}u_i|]dVd\tau.
\end{aligned}
\tag{3.3.42}
$$

Next, we choose the constant γ so large that

$$\gamma + \lambda_3 > 0, \quad 2\gamma - |\lambda_2 - \lambda_3| \geqq 2.$$

We now define the function J on $[0,t_1]$ by

$$J = 2k_1 * D + \gamma[(2 * D)(2 * K)]^{1/2} + k_2 C^2, \tag{3.3.43}$$

where

$$k_1 = (|\lambda_1| + |\lambda_3|)t_1 + 1, \; k_2 = \frac{1}{2}\max[(\gamma+2)t_1, (|\lambda_3|+2)t_1 + 1, 4t_1, 2 + 2|\lambda_3|t_1].$$

Then (3.3.42) implies that

$$y \leqq m * y + J,$$

where $m = \gamma + \lambda_3$. Clearly, $J(s) \leqq J(t)$ if $0 \leqq s \leqq t$. Consider the relation

$$y(s) \leqq m \int_0^s y(z)dz + J(t).$$

This inequality is equivalent to

$$\frac{d}{ds}\left(\exp(-ms)\int_0^s y(z)dz\right) \leqq J(t)\exp(-ms). \tag{3.3.44}$$

It follows from (3.3.44) that

$$y \leqq J\exp(mt_1). \tag{3.3.45}$$

Let

$$r = \{(2 * D)(2 * K) - (1 * G)^2\}^{1/2}. \tag{3.3.46}$$

By (3.3.35),

$$\begin{aligned} &[(2 * D)(2 * K)]^{1/2} = (r^2 + \frac{1}{4}\dot{P}^2)^{1/2} \leqq r + \frac{1}{2}|\dot{P}|, \\ &|\dot{P}| \leqq 2D + 2D(0) = \dot{P} + 4D(0). \end{aligned} \tag{3.3.47}$$

It follows from (3.3.46) and (3.3.47) that

$$J \leqq 2k_1 * D + \gamma r + \frac{1}{2}\gamma\dot{P} + (2\gamma + k_2)C^2. \tag{3.3.48}$$

The relations (3.3.33), (3.3.34), (3.3.42) and (3.3.48) yield

$$y \leqq k_3P + k_4\dot{P} + k_5r, \tag{3.3.49}$$

where

$$k_3 = \max(k_1, 2\gamma + k_2)\exp(mt_1),\ k_5 = 2k_4 = \gamma\exp(mt_1).$$

By (3.3.35), (3.3.36) and (3.3.49),

$$\ddot{P} \geqq 8 * K - 2k_3P - 2k_4\dot{P} - 2k_5r - C^2. \tag{3.3.50}$$

Thus, from (3.3.33)-(3.3.35), (3.3.46) and (3.3.50) we obtain

$$P\ddot{P} - \dot{P}^2 \geqq 4r^2 + 8s^2(1 * K) - 2k_3P^2 - 2k_4P\dot{P} - 2k_5rP -$$

$$- c^2P = (2r - \frac{1}{2} k_5P)^2 + 8s^2(1 * K) - \tag{3.3.51}$$

$$- (2k_3 + \frac{1}{4} k_5^2)P^2 - 2k_4P\dot{P} - c^2P \geqq -m_1P^2 - m_2P\dot{P},$$

where

$$s^2 = 2(t_1 - t)D(0) + c^2, \quad m_1 = 2k_3 + 1 + \frac{1}{4}k_5^2, \quad m_2 = 2k_4.$$

Next, let $\tau = \exp(-m_2t)$. Then we may write (3.3.51) as

$$\frac{d^2}{d\tau^2} \{\ln[P(\tau)\tau^c]\} \geqq 0 \qquad \tau \in (\tau_1, 1),$$

where

$$c = -m_1/m_2^2, \quad \tau_1 = \exp(-m_2t_1).$$

Jensen's inequality together with the continuity of P then gives

$$P(t) \leqq [P(0)]^{\delta}[P(t_1)\exp(m_1t_1/m_2)]^{1-\delta}\exp(-m_1t/m_2), \; t \in (0, t_1) \tag{3.3.52}$$

where δ is given by (3.3.32).

If B_0 is bounded, then we choose R so large that B_0 is contained in the sphere of radius R centered at $\underline{X}_0$. Then, the inequality (3.3.52) coincides with (3.3.31). The inequality (3.3.31), with $\alpha = 0$, implies the continuous dependence of $\underline{u}$ on the initial data and body force. We now assume that B_0 is the exterior of a properly regular region of E^3 whose boundary is a piecewise smooth surface. Letting $R \to \infty$ in (3.3.52) and using (3.3.33), (3.3.34), by the dominated convergence theorem we obtain the inequality (3.3.31). If there exists $\bar{t} \in (0, t_1]$ such that

$$\int_0^{\bar{t}} \int_{B_0} \rho_0 \underline{u}^2 dVdt < \infty, \quad \lim_{\alpha \to 0} \alpha^2 \int_0^{\bar{t}} \int_{B_0} \rho_0^{-1} g S_{Ai}(\underline{u}) S_{Ai}(\underline{u}) dVdt = 0$$

then the inequality (3.3.31) implies again the continuous dependence of $\underline{u}$ on the initial data and body force. □

As a corollary of this theorem we have the following result

Theorem 3.3.3. Let $\underline{v}$ be the difference of two solutions of the problem (A), and assume that the hypotheses of Theorem 3.3.2 hold. If there exists $t' \in (0,t_1]$ such that

$$\int_0^{t'} \int_{B_0} \rho_0 v^2 dVdt < \infty, \quad \lim_{\alpha\to 0} \alpha^2 \int_0^{t'} \int_{B_0} \rho_0^{-1} g S_{Ai}(\underline{v}) S_{Ai}(\underline{v}) dVdt = 0,$$

then $\underline{v} = \underline{0}$ on $B_0 \times [0,t']$.

3.4 Reciprocity. Variational theorem

Throughout this section we assume that the primary state is a configuration of equilibrium. We shall present a reciprocal theorem and a variational theorem of Gurtin type [152] (see also Gurtin [153], Sect. 64).

We first establish a reciprocal theorem by using the method given in [186], [188]. This theorem implies that for null boundary data the differential operator is symmetric in convolution. It is important to note that this fact leads to variational theorems of Gurtin type (see [190]). The reciprocal theorem extends the well-known Graffi's reciprocal theorem (see Gurtin [153], Sect. 61), from classical elasticity.

Consider two external data systems $L^{(\alpha)} = (\underline{F}^{(\alpha)}, \underline{\tilde{u}}^{(\alpha)}, \underline{\tilde{p}}^{(\alpha)}, \underline{\tilde{a}}^{(\alpha)}, \underline{\tilde{b}}^{(\alpha)})$ $(\alpha = 1,2)$ on $\bar{B} \times [0,t_1)$. Let $p^{(\alpha)} = (u_i^{(\alpha)}, s_{rs}^{(\alpha)})$ be the elastic process on $\bar{B} \times [0,t_1)$ corresponding to the external data system $L^{(\alpha)}$. We define

$$h_k^{(\alpha)} = \rho(i * F_k^{(\alpha)} + \tilde{a}_k^{(\alpha)} + t\tilde{b}_k^{(\alpha)}), \quad s_j^{(\alpha)} = s_{kj}^{(\alpha)} n_k, \quad (\alpha = 1,2). \tag{3.4.1}$$

Theorem 3.4.1. Let $p^{(\alpha)}$ be elastic processes corresponding to the external data systems $L^{(\alpha)}$, $(\alpha = 1,2)$. Then

$$\int_B \underline{h}^{(1)} * \underline{u}^{(2)} dv + \int_{\partial B} i * \underline{s}^{(1)} * \underline{u}^{(2)} da = \int_B \underline{h}^{(2)} * \underline{u}^{(1)} dv + \int_{\partial B} i * \underline{s}^{(2)} * \underline{u}^{(1)} da. \tag{3.4.2}$$

Proof. In view of (1.2.45) we find that

$$s_{pq}^{(1)} * u_{q,p}^{(2)} = s_{pq}^{(2)} * u_{q,p}^{(1)} . \tag{3.4.3}$$

If we define

$$I_{\alpha\beta} = \int_B i * s_{pq}^{(\alpha)} * u_{q,p}^{(\beta)} \, dv, \tag{3.4.4}$$

then (3.4.3) implies

$$I_{12} = I_{21}. \tag{3.4.5}$$

By the divergence theorem and Theorem 3.2.2,

$$I_{\alpha\beta} = \int_{\partial B} i*s_j^{(\alpha)}*u_j^{(\beta)} da + \int_B [h_j^{(\alpha)}*u_j^{(\beta)} - \rho u_j^{(\alpha)}*u_j^{(\beta)}]dv. \tag{3.4.6}$$

From (3.4.5) and (3.4.6) we obtain (3.4.2). □

Clearly, the applications of Graffi's theorem (see, for example, Gurtin [153], Sect. 61) can be extended to the present theory.

We note that (3.4.4) and (3.4.6) imply

$$\int_B h_i^{(1)}*u_i^{(1)} dv = \int_B [i*d_{ijrs}u_{r,s}^{(1)} + \rho u_i^{(1)} * u_i^{(1)}]dv - \int_{\partial B} i * s_j^{(1)} * u_j^{(1)} \, da. \tag{3.4.7}$$

Let D denote the set of all admissible displacement fields on $\bar{B} \times [0,t_1)$ which satisfy the boundary conditions

$$u_i = 0 \text{ on } \bar{S}_1 \times [0,t_1), \quad d_{ijrs}u_{r,s}n_j = 0 \text{ on } S_2 \times [0,t_1). \tag{3.4.8}$$

Let $\underline{A}$ be the operator on D defined by

$$A_j\underline{u} = -i * (d_{jkrs}u_{r,s})_{,k} + \rho u_j. \tag{3.4.9}$$

Clearly, in view of Theorem 3.2.3, the mixed problem with null boundary data consists in finding an admissible displacement field $\underline{u}$ on $\bar{B} \times [0,t_1)$ such that $\underline{u} \in D$, and

$$\underline{A}\ \underline{u} = \underline{h} \text{ on } B \times [0,t_1). \tag{3.4.10}$$

For convenience, we introduce the following notation

$$(\underline{f} \circledast \underline{g}) = \int_B f_i * g_i \, dv.$$

It follows from (3.4.7) that

$$(\underline{A}\ \underline{u} \circledast \underline{u}) = \int_B (i * d_{ijrs} u_{r,s} * u_{i,j} + \rho u_i * u_i) dv, \ \underline{u} \in D. \tag{3.4.11}$$

Let $\underline{u}, \underline{v} \in D$. Then (3.4.2) implies

$$(\underline{A}\ \underline{u} \circledast \underline{v}) = (\underline{u} \circledast \underline{A}\ \underline{v}),$$

so that $\underline{A}$ is symmetric in convolution. For each $t \in [0,t_1)$ let $F_t\{\cdot\}$ be the functional on D defined by

$$F_t(\underline{u}) = (\underline{A}\ \underline{u} \circledast \underline{u}) - 2(\underline{u} \circledast \underline{h}).$$

Since $\underline{A}$ is symmetric in convolution, it follows that $\delta F_t\{\underline{u}\} = 0$, $(0 \leqq t < t_1)$, at $\underline{u} \in D$ if and only if $\underline{u}$ is a solution of (3.4.10). In view of (3.4.11), we find that

$$F_t\{\underline{u}\} = \int_B (i * d_{kjrs} u_{r,s} * u_{k,j} + \rho u_j * u_j - 2h_j * u_j) dv,$$

for every $\underline{u} \in D$ and $0 \leqq t < t_1$. Thus, we have established the following result

<u>Theorem 3.4.2.</u> For each $t \in [0,t_1)$ let $F_t\{\cdot\}$ be defined on D by the above relation. Then

$$\delta F_t\{\underline{u}\} = 0, \quad (0 \leqq t < t_1)$$

at $\underline{u} \in D$ if and only if $\underline{u}$ is a solution of the mixed problem with null boundary data.

A direct proof of Theorem 3.4.2 can be given following the proof of the next theorem.

We now state, without proof, two lemmas due to Gurtin [152] (see also Gurtin [153], Sect. 64).

Lemma 3.4.1. Let W be a finite-dimensional inner product space. Let $\underline{f} : B \times [0,t_1) \to W$ be continuous and satisfy

$$\int_B \underline{f} * \underline{g} \, dv = 0 \quad (0 \leq t < t_1)$$

for every class C^∞ function $\underline{g} : B \times [0,t_1) \to W$ that vanishes near ∂B. Then $\underline{f} = \underline{0}$ on $\bar{B} \times [0,t_1)$.

Lemma 3.4.2. Let W be a finite-dimensional inner product space. Let $\underline{g} : S_2 \times [0,t_1) \to W$ be piecewise regular and continuous in time, and assume that

$$\int_{S_2} \underline{f} * \underline{g} \, da = 0 \quad (0 \leq t < t_1)$$

for every class C^∞ function $\underline{g} : \bar{B} \times [0,t_1) \to W$ that vanishes near S_1. Then $\underline{f} = \underline{0}$ on $S_2 \times [0,t_1)$.

The role of these lemmas is analogous to the role of the fundamental lemma in the calculus of variations.

We now present a variational theorem due to Gurtin [152] (see also Gurtin [153], Sect. 64). By a kinematically admissible displacement field we mean an admissible displacement field on $\bar{B} \times [0,t_1)$ which meets the displacement boundary condition.

Theorem 3.4.3. Let M denote the set of all kinematically admissible displacement fields, and for each $t \in [0,t_1)$ define the functional $H_t\{\cdot\}$ on M by

$$H_t(\underline{u}) = \int_B [\frac{1}{2} i * d_{ijrs} u_{r,s} * u_{i,j} + \frac{1}{2} \rho u_s * u_s - h_s * u_s] dv -$$

$$- \int_{S_2} i * \tilde{p}_j * u_j \, da,$$

for every $\underline{u} \in M$. Then

$$\delta H_t\{\underline{u}\} = 0, \qquad (0 \leq t < t_1), \tag{3.4.12}$$

at $\underline{u} \in M$ if and only if $\underline{u}$ is a solution of the mixed problem.

<u>Proof</u>. Let $\underline{u}^o$ be an admissible displacement field on $\bar{B} \times [0,t_1)$, and suppose that

$$\underline{u} + \lambda \underline{u}^o \in M \text{ for every scalar } \lambda. \tag{3.4.13}$$

Condition (3.4.13) is equivalent to the requirement that $\underline{u}^o$ meets the boundary condition

$$\underline{u}^o = 0 \text{ on } \bar{S}_1 \times [0,t_1). \tag{3.4.14}$$

Since

$$\delta_{\underline{u}^o} H_t\{\underline{u}\} = \frac{d}{d\lambda} H_t\{\underline{u} + \lambda \underline{u}^o\}\Big|_{\lambda=0} ,$$

in view of (1.2.45) we arrive at

$$\begin{aligned} \delta_{\underline{u}^o} H_t\{u\} = \int_B [i * d_{ijrs} u_{r,s} * u^o_{i,j} + \rho u_i * u^o_i - h_i * u^o_i] dv - \\ - \int_{S_2} i * \tilde{p}_j * u^o_j \, da, \quad (0 \leq t < t_1), \end{aligned} \tag{3.4.15}$$

for every admissible displacement field $\underline{u}^o$ that satisfies (3.4.14). By the divergence theorem and (3.4.14),

$$\int_B d_{ijrs}u_{r,s} * u^o_{i,j} dv = \int_{S_2} d_{ijrs}u_{r,s}n_j * u^o_i da - \tag{3.4.16}$$
$$- \int_B (d_{ijrs}u_{r,s})_{,j} * u^o_i dv.$$

It follows from (3.4.15) and (3.4.16) that

$$\delta_{\underline{u}^o} H_t\{\underline{u}\} = \int_B [\rho u_i - h_i - i * (d_{ijrs}u_{r,s})_{,j}] * u^o_i dv + \tag{3.4.17}$$
$$+ \int_{S_2} i * (d_{ijrs}u_{r,s}n_j - \tilde{p}_i) * u^o_i da, \quad (0 \leqq t < t_1),$$

for every admissible displacement field $\underline{u}^o$ that satisfies (3.4.14). If $\underline{u}$ is a solution of the mixed problem, then we conclude from (3.4.17) and Theorem 3.2.3 that

$$\delta_{\underline{u}^o} H_t \underline{u} = 0 \quad (0 \leqq t < t_1), \tag{3.4.18}$$

for every admissible displacement field $\underline{u}^o$ that satisfies (3.4.14). This fact yields (3.4.12).

To prove the converse assertion assume that (3.4.12) and hence (3.4.18) holds. Let $\underline{u}^o$ vanish near ∂B. Then it follows from (3.4.17) and (3.4.18) that

$$\int_B [\rho u_i - h_i - i * (d_{ijrs}u_{r,s})_{,j}] * u^o_i dv = 0 \quad (0 \leqq t < t_1), \tag{3.4.19}$$

and (3.4.19) must hold for every $\underline{u}^o$ of class C^∞ on $\bar{B} \times [0,t_1)$ that vanishes near ∂B. But this fact together with Lemma 3.4.1 implies $i*(d_{jkrs}u_{r,s})_{,k} + \rho h_j = \rho u_j$ on $B \times [0,t_1)$. By (3.4.17), (3.4.18) and the result established thus far,

$$\int_{S_2} i * (d_{ijrs}u_{r,s}n_j - \tilde{p}_i) * u^o_i da = 0 \quad (0 \leqq t < t_1),$$

for every admissible displacement field $\underline{u}^o$ that satisfies (3.4.14). Then, Lemma 3.4.2 implies $d_{ijrs}u_{r,s}n_j = \tilde{p}_i$ on $S_2 \times [0,t_1)$. Since $\underline{u} \in M$, Theorem

3.2.3 implies that $\underline{u}$ is a solution of the mixed problem. □

A rational scheme to deduce the functional $H_t\{\cdot\}$ was given in [190].

3.5 Plane harmonic waves in prestressed materials

In this section we study the propagation of a plane wave of small amplitude in an infinite body which is subjected to a static homogeneous deformation. This problem has been investigated in various papers (see, for example, Toupin and Bernstein [350], Hayes and Rivlin [165], Hayes [167], Green [137], Beatty [21], Sawyers and Rivlin [319], [320]).

We assume that the primary state is obtained from B_0 by a finite homogeneous deformation. If $F_i = 0$, then the equations (3.2.10) become

$$d_{ijrs}u_{r,sj} = \rho\ddot{u}_i. \qquad (3.5.1)$$

We seek solutions to (3.5.1) of the form

$$\underline{u} = \underline{a}\,\sin(k\ell_i x_i - \omega t), \quad \underline{\ell}^2 = 1, \qquad (3.5.2)$$

where a_i, k, ℓ_i and ω are constants. A displacement of the form (3.5.2) represents a plane wave of amplitude $\underline{a}$ and wave number k, propagating in the direction $\underline{\ell}$. If we substitute (3.5.2) into (3.5.1) and define the wave velocity by $c = \omega/k$, we obtain

$$(G_{ir} - \rho c^2\,\delta_{ir})a_r = 0, \qquad (3.5.3)$$

where

$$G_{ir} = d_{ijrs}\,\ell_s\ell_j. \qquad (3.5.4)$$

It follows from (1.2.45) and (3.5.4) that

$$G_{ij} = G_{ji}. \qquad (3.5.5)$$

The condition for plane waves is

$$\det(G_{ij} - \rho c^2 \delta_{ij}) = 0. \tag{3.5.6}$$

Since the matrix (G_{ij}) is real and symmetric, the values of c^2 given by (3.5.6) are necessarily real. We now consider another rectangular Cartesian coordinate system $\underline{x}^*$ related to the system $\underline{x}$ by the orthogonal transformation

$$x_i^* = h_{ij}x_j, \quad h_{ik}h_{jk} = h_{ki}h_{kj} = \delta_{ij}, \quad \det(h_{ij}) = 1. \tag{3.5.7}$$

The components u_j^* of the displacement field in the system $\underline{x}^*$ are given by

$$u_j^* = h_{jr}u_r = b_j \sin(k\ell_i x_i - \omega t),$$

where $b_j = h_{jr}a_r$. The equation (3.5.6) becomes

$$\det(M_{ij} - \rho c^2 \delta_{ij}) = 0, \tag{3.5.8}$$

where

$$M_{ij} = h_{ip}h_{jq}G_{pq}. \tag{3.5.9}$$

In view of (3.5.5), there exists a transformation of type (3.5.7) such that $M_{ij} = 0$, $i \neq j$. If we choose (h_{ij}) as the orthogonal matrix representing this transformation, then the wave velocities are

$$c_s^2 = M_{ss}/\rho, \quad \text{(s not summed)}. \tag{3.5.10}$$

The directions of the axes of $\underline{x}^*$ are the three directions of polarisation for which waves may be propagated in the direction of $\underline{\ell}$. If (G_{ij}) is a positive definite matrix, then M_{11}, M_{22} and M_{33} are positive and c_s are real. If one of the solutions for c^2 is negative, then it is easily shown that the primary state is unstable [165]. The necessary and sufficient condition for stability in this case is the positive definiteness of G_{ij}. Thus, the necessary and sufficient conditions for the velocities to be real for arbitrary direction of propagation of the wave, may be expressed as

$$G_{rr} > 0, \quad G_{ij}G_{ij} < G_{rr}G_{ss}, \quad \det(G_{ij}) > 0.$$

These conditions were first presented by Hayes and Rivlin [165]. Certain necessary conditions that waves propagated in an arbitrary direction have real speeds were obtained in [165], [315].

For the remainder of this section we assume that the body B_0 is homogeneous and isotropic. We suppose that the primary state is obtained from B_0 by the pure homogeneous deformation (2.4.1) with extension rations λ_i and principal directions parallel to the axes of the coordinate system. Let us assume that $\underline{\ell}$ is in the direction of the axis x_1. If $\underline{\ell}$ is parallel to a principal axis of strain, then the wave is called a principal wave. It follows from (2.4.5) and (3.5.4) that

$$G_{ij} = 0 \ (i \neq j), \ G_{11} = c_{11} + t_{11}, \ G_{22} = c_{66} + t_{11}, \ G_{33} = c_{55} + t_{11}.$$

We see that there will be three uncoupled principal waves polarized in the principal directions. One of these waves is longitudinal and the other two transverse. The velocity of the longitudinal waves is defined by

$$c_1^2 = (c_{11} + t_{11})/\rho.$$

The velocities of transverse waves polarized in the x_2- and x_3-directions are given by

$$c_2^2 = (c_{66} + t_{11})/\rho, \quad c_3^2 = (c_{55} + t_{11})/\rho,$$

respectively.

These results may be used for the experimental determination of the elastic coefficients. Detailed discussions of this subject may be found in the works of Toupin and Bernstein [350], Thurston and Brugger [344], and Thurston [345].

In [318], Sawyers and Rivlin have obtained, in explicit form, necessary and sufficient conditions that the velocities of sinusoidal waves of infinitesimal amplitude be real for all directions of propagation parallel to a principal plane of the pure homogeneous deformation for an incompressible material. The implications of these results with respect to propagation in an arbitrary direction have been studied in [319]. In what follows we derive, for compressible materials, necessary and sufficient conditions that the wave velocities be real for any direction of propagation parallel to a principal plane of the pure homogeneous deformation (2.4.1). The results we give here are due to Sawyers and Rivlin [320]. We now consider that the direction of propagation of a wave lies in a principal plane, say parallel to the $x_1 0 x_2$-plane. Then, we have $\ell_3 = 0$, and (1.2.44), (2.4.6) and (3.5.4) lead to

$$G_{\alpha\beta} = d_{\alpha\rho\beta\nu}\ell_\rho\ell_\nu, \quad G_{33} = d_{3\alpha 3\beta}\ell_\alpha\ell_\beta, \quad G_{\alpha 3} = 0. \tag{3.5.11}$$

We note that

$$\begin{aligned}
&d_{1111} = c_{11} + t_{11} = c_{11} + \lambda_1^2 K_3 + \lambda_1^2\lambda_2^2 L_3, \\
&d_{2222} = c_{22} + t_{22} = c_{22} + \lambda_2^2 K_3 + \lambda_1^2\ \lambda_2^2 L_3, \\
&d_{1122} = c_{12}, \quad d_{1221} = c_{66} = -\lambda_1^2\lambda_2^2 L_3, \\
&d_{1212} = c_{66} + t_{22} = \lambda_2^2 K_3, \quad d_{2121} = c_{66} + t_{11} = \lambda_1^2 K_3, \\
&d_{3131} = c_{55} + t_{11} = \lambda_1^2 K_2, \quad d_{3232} = c_{44} + t_{22} = \lambda_2^2 K_1,
\end{aligned} \tag{3.5.12}$$

where

$$K_i = \frac{2}{\sqrt{I_3}}\left(\frac{\partial e}{\partial I_1} + \lambda_i^2\,\frac{\partial e}{\partial I_2}\right), \quad L_i = \frac{2}{\sqrt{I_3}}\left(\frac{\partial e}{\partial I_2} + \lambda_i^2\,\frac{\partial e}{\partial I_3}\right). \tag{3.5.13}$$

It follows from (3.5.11) and (3.5.12) that

$$\begin{aligned}
&G_{11} = (c_{11}\lambda_1^{-2} + K_3 + \lambda_2^2 L_3)\lambda_1^2\ell_1^2 + \lambda_2^2\ell_2^2 K_3, \\
&G_{22} = (c_{22}\lambda_2^{-2} + K_3 + \lambda_1^2 L_3)\lambda_2^2\ell_2^2 + \lambda_1^2\ell_1^2 K_3, \\
&G_{33} = \lambda_1^2\ell_1^2 K_2 + \lambda_2^2\ell_2^2 K_1, \ G_{12} = (\bar{c}_{12} + \lambda_1^2\lambda_2^2 L_3)\ell_1\ell_2,
\end{aligned} \tag{3.5.14}$$

where

$$\bar{c}_{12} = c_{12} + 2c_{66}. \tag{3.5.15}$$

The equation (3.5.6) yields

$$\rho c^2 = G_{33}, \tag{3.5.16}$$

or

$$(\rho c^2 - G_{11})(\rho c^2 - G_{22}) - G_{12}^2 = 0. \tag{3.5.17}$$

In view of (3.5.13), the necessary and sufficient conditions that (3.5.16) yield only positive values for c^2 for all ℓ_1 and ℓ_2 are $K_\alpha > 0$. The study of waves propagated parallel to $x_2 0 x_3$ or $x_1 0 x_3$ plane leads additionally to the condition $K_3 > 0$. We accordingly have the conditions

$$K_i > 0. \tag{3.5.18}$$

We note that (3.5.18) are the well-known Baker-Ericksen [18] conditions for stability.

The necessary and sufficient conditions that (3.5.17) yield only positive values for c^2 are

$$G_{11} + G_{22} > 0, \quad G_{11}G_{22} - G_{12}^2 > 0. \tag{3.5.19}$$

It follows from (3.5.14) that

$$\begin{aligned} G_{11} + G_{22} &= (F_1 + K_3)\lambda_1^2\ell_1^2 + (F_2 + K_3)\lambda_2^2\ell_2^2, \\ G_{11}G_{22} - G_{12}^2 &= (\lambda_1^4\ell_1^4 F_1 + \lambda_2^4\ell_2^4 F_2)K_3 + 2\lambda_1^2\lambda_2^2\ell_1^2\ell_2^2 M, \end{aligned} \tag{3.5.20}$$

where

$$\begin{aligned} F_1 &= c_{11}\lambda_1^{-2} + K_3 + \lambda_2^2 L_3, \quad F_2 = c_{22}\lambda_2^{-2} + K_3 + \lambda_1^2 L_3, \\ 2M &= F_1F_2 + K_3^2 - \lambda_1^2\lambda_2^2(L_3 + \bar{c}_{12}\lambda_1^{-2}\lambda_2^{-2})^2. \end{aligned} \tag{3.5.21}$$

We note that F_α can be rewritten as the first two of the relations

$$F_i = c_{ii}\lambda_i^{-2} + \frac{2}{\sqrt{I_3}}[\frac{\partial e}{\partial I_1} + (I_1 - \lambda_i^2)\frac{\partial e}{\partial I_2} + I_3\lambda_i^{-2}\frac{\partial e}{\partial I_3}]. \qquad (3.5.22)$$

The necessary and sufficient conditions for $(3.5.19)_1$ to be satisfied for all ℓ_1 and ℓ_2 are $F_1 + K_3 > 0$ and $F_2 + K_3 > 0$. In an analogous manner, by considering waves to be propagated parallel to the $x_2 0 x_3$ or $x_1 0 x_3$ planes we find additionally the condition $F_3 + K_3 > 0$. Thus, we have the conditions

$$F_i + K_i > 0. \qquad (3.5.23)$$

The necessary and sufficient conditions that the inequality $(3.5.19)_2$ be satisfied for all ℓ_1, ℓ_2 are the first two of the conditions

$$F_i > 0. \qquad (3.5.24)$$

Clearly, (3.5.18) and (3.5.24) imply (3.5.23). We have

$$G_{11}G_{22} - G_{12}^2 = K_3[(\lambda_1^2\ell_1^2F_1^{1/2} - \lambda_2^2\ell_2^2F_2^{1/2})^2 + + 2\lambda_1^2\lambda_2^2\ell_1^2\ell_2^2(F_1^{1/2}F_2^{1/2} + MK_3^{-1})]. \qquad (3.5.25)$$

It follows that, provided the conditions (3.5.18) and (3.5.24) are satisfied, the necessary and sufficient condition for $(3.5.19)_2$ to be satisfied is

$$K_3(F_1F_2)^{1/2} + M > 0. \qquad (3.5.26)$$

In view of (3.5.21), this condition can be expressed as the first of the conditions

$$(F_iF_r)^{1/2} + K_s > \lambda_i\lambda_r|L_s + c_{ir}(\lambda_i\lambda_r)^{-2}|$$
$$(i,r,s) \in \{(1,2,3),(2,3,1),(3,1,2)\}. \qquad (3.5.27)$$

The third of the conditions (3.5.24), and the second and third of the conditions (3.5.27) may be obtained in an analogous manner by considering waves to be propagated parallel to the $x_1 0 x_3$ and $x_2 0 x_3$ planes.

Thus, the necessary and sufficient conditions that waves propagated parallel to a principal plane of the pure homogeneous deformation shall have real velocities are given by (3.5.18), (3.5.24) and (3.5.27).

3.6 Surface waves

In this section we study the propagation of surface waves in a semi-infinite body which is subjected to a static homogeneous deformation. The results we give here are due to Hayes and Rivlin [166].

We assume that the body B_0 occupies the half space $X_2 \geqq 0$. This body is subjected to the homogeneous deformation defined by (2.4.1). We suppose that B_0 is homogeneous and isotropic. Then, the stress components are given by (2.4.3). We assume that in the primary deformation the boundary of B_0 is free of tractions. Thus we obtain the condition

$$t_{22} = 0. \tag{3.6.1}$$

Then, we have

$$p = -\lambda_2^2\Phi - \lambda_2^2(\lambda_1^2 + \lambda_3^2)\Psi. \tag{3.6.2}$$

It follows from (2.4.3) and (3.6.2) that

$$\begin{gathered} t_{11} = (\lambda_1^2 - \lambda_2^2)(\Phi + \lambda_3^2\Psi), \quad t_{33} = (\lambda_3^2 - \lambda_2^2)(\Phi + \lambda_1^2\Psi), \\ t_{22} = 0, \; t_{ij} = 0 \qquad (i \neq j). \end{gathered} \tag{3.6.3}$$

We superimpose on the deformation (2.4.1) an infinitesimal displacement $\underline{u}$. We assume that the body forces are zero. Substitution of (2.4.5) into the equations (3.2.7) yields the equations

$$\begin{aligned} &(c_{11} + t_{11})u_{1,11} + c_{66}u_{1,22} + (c_{55} + t_{33})u_{1,33} + \\ &\quad + (c_{12} + c_{66})u_{2,12} + (c_{13} + c_{55})u_{3,13} = \rho\ddot{u}_1, \end{aligned}$$

$$(c_{12} + c_{66})u_{1,12} + (c_{66} + t_{11})u_{2,11} + c_{22}u_{2,22} + $$
$$+ (c_{44} + t_{33})u_{2,33} + (c_{23} + c_{44})u_{3,23} = \rho\ddot{u}_2, \tag{3.6.4}$$

$$(c_{13} + c_{55})u_{1,13} + (c_{23} + c_{44})u_{2,23} + (c_{55} + t_{11})u_{3,11} +$$
$$+ c_{44}u_{3,22} + (c_{33} + t_{33})u_{3,33} = \rho\ddot{u}_3.$$

We assume that in the secondary deformation the boundary $x_2 = 0$ is a free surface. Hence the boundary conditions on $x_2 = 0$ are

$$s_{2i} = 0. \tag{3.6.5}$$

We look for a solution representing a plane wave propagating in a direction parallel to the surface of the half space. This solution if it exists, is known as the surface wave. In the classical theory of elastic solids it was first discovered by Rayleigh.

We suppose that the displacement vector field has the form

$$u_\alpha = \mathrm{Re}\{f_\alpha(x_2)\exp i(kx_1 + qt)\}, \; u_3 = 0, \; (i = \sqrt{-1}), \tag{3.6.6}$$

where Re { } denotes the real part of the quantity { } , and f_α are to be determined.

It follows from (3.6.4) and (3.6.6) that

$$[\rho q^2 - k^2(c_{11}+t_{11})]f_1 + c_{66}f_1'' + ik(c_{12} + c_{66})f_2' = 0,$$
$$[\rho q^2 - k^2(c_{66} + t_{11})]f_2 + c_{22}f_2'' + ik(c_{12} + c_{66})f_1' = 0, \tag{3.6.7}$$

where $f' = df/dx_2$. If we take

$$f_1 = -ik(c_{12}+c_{66})g', \; f_2 = [\rho q^2-k^2(c_{11}+t_{11})]g + c_{66}g'', \tag{3.6.8}$$

where g is an unknown function, then the first of equations (3.6.7) is satisfied identically. From the second of (3.6.7) we obtain

$$c_{22}c_{66}g^{(4)} + \rho q^2(c_{22} + c_{66} - dw)g'' + \rho^2 q^4[1 - (c_{11}+t_{11})w][1 - (c_{66}+t_{11})w] = 0, \tag{3.6.9}$$

where

$$w = k^2/(\rho q^2),\ d = (c_{12}+c_{66})^2 - (c_{11}+t_{11})c_{22} - (c_{66}+t_{11})c_{66}. \tag{3.6.10}$$

The general solution of the equation (3.6.9) is

$$g = \sum_{\alpha=1}^{2} [A_\alpha \exp(-m_\alpha x_2) + B_\alpha \exp(m_\alpha x_2)],$$

where A_α and B_α are constants and $\pm\, m_\alpha$ are the roots of the equation

$$c_{22}c_{66}m^4 + \rho q^2(c_{22} + c_{66} - dw)m^2 + \rho^2 q^4[1 - (c_{11}+t_{11})w][1 - (c_{66}+t_{11})w] = 0. \tag{3.6.11}$$

Since u_α are to tend to zero as x_2 tends to infinity, then, taking Re $m_\alpha > 0$, we must have $B_\alpha = 0$. Then

$$\begin{aligned} u_1 &= \mathrm{Re}\ \{ik(c_{12}+c_{66}) \sum_{\alpha=1}^{2} m_\alpha A_\alpha \exp[-m_\alpha x_2+i(kx_1 + qt)]\}, \\ u_2 &= \mathrm{Re}\ \{\sum_{\alpha=1}^{2} [\rho q^2-k^2(c_{11}+t_{11})+c_{66}m_\alpha^2]A_\alpha \exp[-m_\alpha x_2+i(kx_1 + qt)]\}. \end{aligned} \tag{3.6.12}$$

The displacements (3.6.12) correspond to waves, propagated parallel to the surface of the body, with their amplitudes decreasing with distance from the surface. In view of (2.4.5) and (3.6.6), the boundary conditions (3.6.5) reduce to

$$c_{12}u_{1,1} + c_{22}u_{2,2} = 0,\ u_{1,2} + u_{2,1} = 0 \text{ on } x_2 = 0. \tag{3.6.13}$$

It follows from (3.6.12) and (3.6.13) that

$$\sum_{\alpha=1}^{2} m_\alpha\{c_{12}(c_{12}+c_{66})k^2+c_{22}[\rho q^2-(c_{11}+t_{11})k^2+c_{66}m_\alpha^2]\}A_\alpha = 0,$$
$$k\sum_{\alpha=1}^{2} [\rho q^2-(c_{11}+t_{11})k^2-c_{12}m_\alpha^2]A_\alpha = 0. \tag{3.6.14}$$

In order that this system have non-vanishing solutions for constants A_α we must have

$$k(m_2-m_1)[P_1P_2 + c_{22}c_{66}P_2(m_1^2 + m_2^2) +$$
$$+ (c_{12}P_1-c_{22}c_{66}P_2)m_1m_2 - c_{22}c_{66}c_{12}m_1^2m_2^2] = 0, \tag{3.6.15}$$

where

$$P_1 = c_{22}\ \rho q^2 + [c_{12}(c_{12}+c_{66}) - c_{22}(c_{11}+t_{11})]k^2,$$
$$P_2 = \rho q^2 - (c_{11} + t_{11})k^2.$$

We find either

$$m_1 = m_2, \quad k = 0, \tag{3.6.16}$$

or

$$P_1P_2 + c_{22}c_{66}P_2(m_1^2+m_2^2) + (c_{12}P_1-c_{22}c_{66}P_2)m_1m_2 -$$
$$- c_{22}c_{66}c_{12}m_1^2m_2^2 = 0. \tag{3.6.17}$$

It follows from (3.6.11) that

$$m_1^2 + m_2^2 = -\rho q^2(c_{22} + c_{66} - dw)/c_{22}c_{66},$$
$$m_1^2m_2^2 = \rho^2q^4[1-(c_{11}+t_{11})w][1 - (c_{66}+t_{11})w]/c_{22}c_{66}. \tag{3.6.18}$$

If we eliminate m_α between (3.6.17) and (3.6.18) we find an equation which implies either

$$(c_{11} + t_{11})w = 1 \tag{3.6.19}$$

or

$$c_{22}c_{66}A(c_{12}B-dw-C + c_{22} + c_{66})^2 = (c_{22}c_{66}A-c_{12}C)^2B, \tag{3.6.20}$$

where

$$A = 1 - (c_{11} + t_{11})w, \quad B = 1 - (c_{66} + t_{11})w,$$

$$C = c_{22} + [c_{12}(c_{12}+c_{66}) - c_{22}(c_{11} + t_{11})]w.$$

It has been shown by Hayes and Rivlin [166] that the solutions corresponding to (3.6.16) and (3.6.19) are degenerate. Thus we call (3.6.20) the frequency equation determining the velocity of surface waves propagating in a homogeneously deformed elastic half space. Since the coefficients of w in (3.6.20) do not involve either k or q, there is no dispersion. In [166], Hayes and Rivlin have derived the changes in the velocities of propagation of surface waves, which are produced by an initial small homogeneous deformation. Moreover, they have specialized the analysis to incompressible bodies, and also studied Love waves.

The problem of determining conditions on the primary deformation which are necessary and sufficient for the existence of a surface wave was considered by Flavin [124] who studied incompressible elastic materials of the Mooney and neo-Hookean types. In [381]-[383], Willson has investigated the properties of surface waves for a variety of isotropic elastic materials, compressible and incompressible, and for different primary deformations. A systematic treatment of the surface waves in homogeneous and anisotropic elastic bodies subjected to a homogeneous deformation has been presented by Chadwick and Jarvis [61]. The propagation of Love waves in prestressed elastic bodies has been discussed by Guz' [157], Babich, Guz' and Zhuk [11], Pan and Chakraborty [283], Kar and Pal [217].

We now give a brief review of other papers concerned with the vibrations of prestressed elastic bodies. Green [136] has investigated the torsional vibrations of an infinite circular cylinder subject to finite longitudinal extension. A solution for longitudinal oscillations of an initially stretched circular cylinder has been found by Şuhubi [336] (see, also Zorski [389]). Demiray and Şuhubi [96] have investigated the vibrations of an initially twisted circular rubber cylinder. Guo [145], [146] has studied the non-rotational vibrations and stability of an incompressible circular cylinder of finite length, extended and inflated initially. Eason [110] has investigated the wave propagation in an elastic rod subjected to initial finite extension and twist. Other results concerning the wave propagation in prestressed elastic cylinders are given by Belward [29] and Moiseenko [264]. Some special motions of an infinite isotropic solid have been considered by Ramakanth [303]. The vibrations of a radially prestressed sphere have been discussed in [304]. Other special problems concerning the vibrations of initially stressed elastic bodies have been studied by Zhuk [388], Benveniste [30], Kalinchuk and Poliakova [214], Dey and Mukherjee [100], Sidhu and Singh [327], Chattopadhyay and Keshri [67].

3.7 Propagation of acceleration waves in homogeneously deformed bodies

The propagation of singular surfaces through a finitely strained material has been studied extensively (see, for example, Truesdell and Noll [356], Chen [74], McCarthy [257], Eringen and Şuhubi [116], Wesolowski [375]).

In this section we present briefly some results concerning the propagation of acceleration waves in homogeneously deformed elastic materials. We describe relations of the acceleration waves to the plane waves studied in Section 3.5.

α) Preliminaries. Let S(t) be a moving surface which admits the representation

$$X_A = \phi_A(\xi^\alpha, t), \tag{3.7.1}$$

where the parameter pairs ξ^α belong to the open subset D of R^2, and $t \in I \subset R$. We shall refer to S(t) as a smooth surface if the functions ϕ_i are one-one and of class C^2 on $D \times I$ and the matrix $(\partial\phi_i/\partial\xi^\alpha)$ is, at all points of S(t), of rank 2.

We assume that S(t) is a smooth surface which intersects the domain B_0. We denote by $B_0^+(t)$ and $B_0^-(t)$ the disjoint open subsets of B_0 whose union is $B_0 \setminus S(t)$. The unit normal $\underline{N}$ of the surface S(t) is directed towards $B_0^+(t)$.

Let $g(\cdot,\cdot)$ be a function such that $g(\cdot,t)$ is continuous within the regions $B_0^+(t)$ and $B_0^-(t)$, and let $g(\underline{x},t)$ have definite limits g^+ and g^- as $\underline{x}$ approaches a point on the surface from $B_0^+(t)$ and $B_0^-(t)$, respectively. The surface $S(t)$ is said to be singular with respect to g at time t if $[g] \equiv g^+ - g^- \neq 0$.

The velocity $\underline{V}$ of a point on the surface is defined by

$$V_A = \dot{\phi}_A. \tag{3.7.2}$$

It follows from (3.7.1) that the surface $S(t)$ may be represented by an equation of the form

$$F(\underline{X},t) = 0, \tag{3.7.3}$$

where the function F is of class C^2 on its domain. In terms of this representation,

$$N_A = MF_{,A}, \tag{3.7.4}$$

where

$$M = (F_{,A}F_{,A})^{-1/2}. \tag{3.7.5}$$

The spped of propagation U_N of the surface $S(t)$ is defined by

$$U_N = \underline{V}\cdot\underline{N} = -M\,\frac{\partial F}{\partial t}. \tag{3.7.6}$$

A singular surface is said to be a wave if U_N is different from zero. The spatial description of the surface $S(t)$ is provided by

$$f(\underline{x},t) = F(\underline{X}^{(-1)}(\underline{x},t),t) = 0. \tag{3.7.7}$$

Equation (3.7.7) represents in the spatial coordinates a moving surface $\sigma(t)$. The alternative forms (3.7.3) and (3.7.7) represent the same moving surface. The unit normal $\underline{n}$ to the surface $\sigma(t)$ is given by

$$n_i = mf_{,i}, \quad m = (f_{,i}f_{,i})^{-1/2}. \tag{3.7.8}$$

The normal speed of the surface $\sigma(t)$, or the speed of displacement, is

$$u_n = - m \frac{\partial f}{\partial t}. \tag{3.7.9}$$

It follows from (3.7.4), (3.7.7) and (3.7.8) that

$$N_K = C_n^{-1} n_i x_{i,K}, \tag{3.7.10}$$

where

$$C_n^2 = x_{j,L} x_{r,L} n_j n_r. \tag{3.7.11}$$

The quantity

$$U = u_n - \dot{x}_i n_i, \tag{3.7.12}$$

is called the local speed of propagation. It can be readily verified that

$$U_N = U C_n^{-1}. \tag{3.7.13}$$

The metric tensor on S(t) is defined by

$$A_{\alpha\beta} = \phi_{K,\alpha} \phi_{K,\beta}.$$

The elements $A^{\alpha\beta}$ are characterized by

$$A^{\alpha\beta} A_{\beta\rho} = \delta^{\alpha}_{\rho} .$$

The second fundamental form on S(t) is given by

$$B_{\alpha\beta} = N_K \phi_{K;\alpha\beta} = - N_{K;\alpha} \phi_{K;\beta},$$

where a semicolon is used to indicate covariant derivatives in the ξ coordinate system.

We shall need the following compatibility conditions at the moving surface S(t) (see Thomas [343], Truesdell and Toupin [354], Sects. 175, 176, 180, 181)

$$[g_{,K}] = \bar{B}N_K + A^{\alpha\beta} A_{;\beta} \phi_{K;\beta},$$

$$[g_{,KL}] = CN_KN_L + A^{\alpha\beta} (\bar{B}_{;\alpha} + A^{\nu\rho} B_{\alpha\nu} A_{;\rho})(N_K \phi_{L;\beta} +$$

$$+ N_L \phi_{K;\beta}) + A^{\alpha\beta} A^{\nu\rho} (A_{;\alpha\nu} - B_{\alpha\nu}\bar{B})\phi_{K;\beta}\phi_{L;\rho},$$

$$[\dot{g}] = \frac{\delta_D A}{\delta t} - U_N\bar{B}, \tag{3.7.14}$$

$$[\dot{g}_{,K}] = A^{\alpha\beta} (\frac{\delta_D A}{\delta t} - U_N\bar{B})_{;\alpha} \phi_{K;\beta} + (\frac{\delta_D \bar{B}}{\delta t} + A^{\alpha\beta} A_{;\alpha}U_{N;\beta} -$$

$$- U_NC)N_K,$$

$$[\ddot{g}] = \frac{\delta_D}{\delta t} (\frac{\delta_D A}{\delta t} - U_N\bar{B}) - U_N(\frac{\delta_D \bar{B}}{\delta t} + A^{\alpha\beta} A_{;\alpha}U_{N;\beta} - U_NC),$$

where

$$A = [g], \quad \bar{B} = [N_Kg_{,K}], \quad C = [g_{,KL}N_KN_L], \tag{3.7.15}$$

and

$$\frac{\delta_D}{\delta t} = \frac{\partial}{\partial t} + U_N \frac{\partial}{\partial N}, \tag{3.7.16}$$

is the displacement derivative with respect to the surface S(t).

A propagating singular surface is said to be an acceleration wave if $\underline{x}(\underline{X},t)$, $\dot{\underline{x}}(\underline{X},t)$ and $\underline{x}_{,K}(\underline{X},t)$, $(\underline{X},t) \in B_0 \times I$ are continuous functions everywhere, and $\ddot{\underline{x}}$, $\underline{x}_{,KL}$, $\dot{\underline{x}}_{,K}$, $\underline{x}_{,KLM}$, $\dot{\underline{x}}_{,LM}$, $\ddot{\underline{x}}_{,K}$, $\dddot{\underline{x}}$ have jump discontinuities across S(t) but are continuous functions everywhere else.

Let

$$\underline{a} = [\ddot{\underline{x}}], \tag{3.7.17}$$

be the jump in the acceleration. The vector $\underline{a}$ is called the amplitude of the wave. It follows from the compatibility conditions on S(t) that for an acceleration wave we have

$$[x_{i,KL}] = \frac{1}{U_N^2} a_i N_K N_L, \quad [\dot{x}_{i,K}] = -\frac{1}{U_N} a_i N_K. \tag{3.7.18}$$

If the amplitude vector $\underline{a}$ is parallel to the normal $\underline{n}$ of the wave surface, we have a longitudinal wave; if it is perpendicular to $\underline{n}$, we have a transverse wave. For longitudinal wave we write

$$\underline{a} = a\underline{n}. \tag{3.7.19}$$

A longitudinal wave for which $a > 0$ is called expansive; a longitudinal wave for which $a < 0$ is called compressive.

In the following we shall study the acceleration waves propagating into an initially homogeneously deformed elastic material. This problem has been studied in various papers (see, for example, Green [143], Chen [72], [74], Şuhubi [337]).

Throughout this section we assume that the body B_o is homogeneous.

We consider the equations of motion in the form (1.1.6). In view of the constitutive equations, the equations of motion become

$$A_{MiKj} x_{j,MK} + \rho_o f_i = \rho_o \ddot{x}_i, \tag{3.7.20}$$

where

$$A_{MiNj} = \frac{\partial^2 e}{\partial x_{i,M} \partial x_{j,N}}. \tag{3.7.21}$$

If we differentiate the equations (3.7.20) with respect to time, we obtain

$$B_{AiKjRs} \dot{x}_{s,R} x_{j,AK} + A_{AiKj} \dot{x}_{j,AK} + \rho_o \dot{f}_i = \rho_o \dddot{x}_i, \tag{3.7.22}$$

where

$$B_{MiKjRs} = \frac{\partial^3 e}{\partial x_{i,M} \partial x_{j,K} \partial x_{s,R}}, \tag{3.7.23}$$

is the second-order elasticity tensor. We assume that the elasticity tensor A_{MiNj} is of class C^1 and the second-order elasticity tensor is of class C^o. Moreover we suppose that the body force is of class C^1.

Consider a surface $\sigma(t)$ propagating into the body B. If $\sigma(t)$ is an acceleration wave, then we have

$$A_{MiKj}[x_{j,MK}] = \rho_0[\ddot{x}_i], \tag{3.7.24}$$

on S(t). It follows from (3.7.18) and (3.7.24) that the amplitude vector $\underline{a}$ and the speed of propagation U_N obey the following propagation condition

$$(Q_{ij}(\underline{N}) - \rho_0 U_N^2 \, \delta_{ij})a_j = 0, \tag{3.7.25}$$

where $Q_{ij}(\underline{N})$ is a symmetric tensor defined by

$$Q_{ij}(\underline{N}) = A_{MiKj}N_M N_K. \tag{3.7.26}$$

Clearly, $Q_{ij}(\underline{N})$ are functions of deformation gradient. $Q_{ij}(N)$ is called the acoustic tensor. It follows that $\rho_0 U_N^2$ is a proper value and the amplitude vector $\underline{a}$ is a proper vector of the tensor $Q_{ij}(\underline{N})$.

Consider a fixed deformation gradient at a point in the body B. Now, the acoustic tensor depends on the normal $\underline{N}$ alone. In view of (3.7.10) we can write

$$Q_{ij}(\underline{N}) = \tilde{Q}_{ij}(\underline{n}). \tag{3.7.27}$$

β) Longitudinal waves. It is known that at a point of a deformed elastic body where

$$\tilde{Q}_{ij}(\underline{n})n_i n_j > 0, \tag{3.7.28}$$

for all unit vectors $\underline{n}$, there is at least one direction in which a longitudinal wave may exist and propagate (see Truesdell [357]). In what follows we assume that since time t = 0, the longitudinal wave has been propagating into a region B which is obtained from B_0 by the homogeneous deformation

$$x_i = \lambda_{iK}X_K, \tag{3.7.29}$$

where λ_{iK} are given constants. In this case the acoustic tensor and the second-order elasticity tensor are constants. We assume that the condition (3.7.28) holds. Then a plane longitudinal wave may exist and propagate in the same direction throughout the body. Let us study the behaviour of this wave. The results we give here are due to Chen [72] (see also Chen [74], Sect. 6).

We denote by $\underline{n}^{(1)}$ the constant unit vector of the direction in which the longitudinal wave may exist and propagate. The vector $\underline{N}^{(1)}$ corresponding to $\underline{n}^{(1)}$ is given by

$$N_K^{(1)} = (\lambda_{jL}\lambda_{rL}n_j^{(1)}n_r^{(1)})^{-1/2}\lambda_{iK}n_i^{(1)}. \tag{3.7.30}$$

The amplitude vector of the longitudinal wave has the form $\underline{a} = a\underline{n}^{(1)}$. The speed of propagation is given by

$$U^2_{N^{(1)}} = \rho_0^{-1}Q_{ij}(\underline{N}^{(1)};L)n_i^{(1)}n_j^{(1)} = \rho_0^{-1}\tilde{Q}_{ij}(\underline{n}^{(1)};L)n_i^{(1)}n_j^{(1)}, \tag{3.7.31}$$

where we have indicated that the acoustic tensor is calculated for the constant deformation gradient $L = (\lambda_{iK})$. We introduce the notations

$$M_{ij} = Q_{ij}(\underline{N}^{(1)};L), \quad U_0 = U_{N^{(1)}}. \tag{3.7.32}$$

We assume that the amplitude a is a function of time only, independent of the parameter pairs ξ^α.

In order to derive the growth equation we evaluate the jumps of the equations (3.7.22) on the wave fron S(t). We obtain

$$B_{MiKjRs}[\dot{x}_{s,R}x_{j,MK}] + \dot{A}_{MiKj}[x_{j,MK}] = \rho_0[\dddot{x}_i]. \tag{3.7.33}$$

The jump in the third-order derivatives may be evaluated from (3.7.14), by recalling that $\dot{x}_i$ and $x_{i,K}$ are continuous across S(t). Thus, we get

$$[\dot{x}_{i,KL}] = c_iN_KN_L, \quad [\dddot{x}_i] = c_iU_N^2 + 2\frac{\delta_D a_i}{\delta t}, \tag{3.7.34}$$

where

$$c_i = [\dot{x}_{i,KL}N_KN_L]. \tag{3.7.35}$$

The vector $\underline{c}$ is called the induced discontinuity associated with the acceleration wave. If we use the relation

$$[fg] = f^+[g] + g^+[f] - [f][g],$$

and the assumption that the material ahead of the wave front is at rest, then we obtain

$$[\dot{x}_{s,R}x_{j,MK}] = -[\dot{x}_{s,R}][x_{j,MK}].$$

It follows from (3.7.18), (3.7.32), (3.7.33) and (3.7.44) that

$$2\,\rho_o \frac{\delta_D a}{\delta t} n_i^{(1)} - \frac{1}{U_o^3} b_i a^2 = (M_{ij} - \rho_o U_o^2\,\delta_{ij})c_j, \tag{3.7.36}$$

where $\underline{b}$ is a constant vector defined by

$$b_i = B_{MiKjRp}(L)N_M^{(1)}N_K^{(1)}N_R^{(1)}n_j^{(1)}n_p^{(1)}. \tag{3.7.37}$$

Since the equation

$$(M_{ij} - \rho_o U_o^2\,\delta_{ij})\alpha_j = 0,$$

has a non-trivial solution it follows that if a solution $\underline{c}$ of the equation (3.7.36) exists, it cannot be unique. We assume first that $\underline{c}$ is a proper vector of the tensor M_{ij} corresponding to the proper value $\rho_o U_o^2$. Then, from (3.7.36) we obtain the equation

$$\frac{\delta_D a}{\delta t} = \frac{\gamma}{2\rho_o U_o^3} a^2, \tag{3.7.38}$$

where

$$\gamma = b_i n_i^{(1)}. \tag{3.7.39}$$

It follows from (3.7.38) that

$$a(t) = a(0)\left(1 - \frac{\gamma a(0)}{2\rho_o U_o^3} t\right)^{-1}. \tag{3.7.40}$$

From this result we conclude that the growth or decay of the acceleration discontinuity is administered by the sign of the constant γ. If $\gamma > 0$, then the amplitude of a compressive wave (i.e. $a(0) < 0$) will decay to zero monotonically in infinite time whereas the amplitude of an expansive wave ($a(0) > 0$) will grow steadily, and after a finite time lapse

$$t_* = \frac{2\rho_o U_o^3}{\gamma a(0)},$$

it will become infinite.

If $\gamma < 0$, the behaviour of expansive and compressive waves are reversed. If $\gamma = 0$, then the amplitude of the wave remains unchanged as it travels in the material. Thus, the criteria governing the growth and decay of the amplitude depend only on the second-order elasticity tensor.

We assume now that the induced discontinuity vector $\underline{c}$ is not a proper vector of the tensor M_{ij} corresponding to the proper value $\rho_o U_o^2$. Multiplying both sides of (3.7.36) by $n_i^{(1)}$, we obtain for the amplitude a the equation (3.7.38). Thus, in this case the solution is given by (3.7.40) and the criteria governing the growth and decay of the amplitude are identical to those given above.

Let us note that the explicit time dependence of certain components of the induced discontinuity vector can be determined. The tensor M_{ij} has two other mutually orthogonal principal axes, say $\underline{n}^{(2)}$ and $\underline{n}^{(3)}$, also orthogonal to $\underline{n}^{(1)}$. It follows from (3.7.36) that

$$b_i n_i^{(g)} a^2 = - U_o^3 (M_{ij} n_i^{(g)} c_j - \rho_o U_o^2 n_r^{(g)} c_r), \quad (g = 2,3).$$

Let $\lambda^{(1)}$ and $\lambda^{(2)}$ be the proper values corresponding to $\underline{n}^{(1)}$ and $\underline{n}^{(2)}$, respectively. Then

$$M_{ij} n_j^{(g)} = \lambda^{(g)} n_i^{(g)} \quad (g = 2,3),$$

so that

$$c_i n_i^{(g)} = b_j n_j^{(g)} a^2/(\rho_0 U_o^2 - \lambda^{(g)}) U_o^3 \quad (g = 2,3),$$

provided $\lambda^{(g)} \neq \rho_0 U_o^2$. Thus, the component of $\underline{c}$ in the $\underline{n}^{(g)}$ direction will either grow or decay according as the amplitude of the primary wave grows or decays.

γ) Transverse waves. It has been proved by Truesdell [357] that at a point of a deformed body such that the material has strongly elliptic elasticity, there is at least one direction in which a longitudinal wave and two transverse waves with orthogonal amplitude vectors may exist and propagate. We now study the behaviour of the transverse waves which propagate into the body B, obtained from B_o by the homogeneous deformation (3.7.29). The results presented here are due to Chen [73].

We assume that the elasticity tensor is strongly elliptic. Let $\underline{n}^{(1)}$ denote the constant unit vector of the direction in which the longitudinal wave may exist and propagate. Let $\underline{N}^{(1)}$ be defined by (3.7.30). Clearly, in this case the transverse waves are necessarily plane and their amplitude vectors are parallel to $\underline{n}^{(2)}$ and $\underline{n}^{(3)}$, the two other principal axes of the tensor M_{ij}. We denote by $\underline{a}^{(2)}$ and $\underline{a}^{(3)}$ the amplitude vectors of the transverse waves

$$\underline{a}^{(g)} = a^{(g)} \underline{n}^{(g)} \qquad (g = 2,3), \tag{3.7.41}$$

with $a^{(g)} > 0$. The corresponding wave speeds will be denoted by U_2 and U_3, respectively. We have

$$\rho_0 U_g^2 = \lambda^{(g)} = Q_{ij}(\underline{N}^{(1)};L) n_i^{(g)} n_j^{(g)}, \quad (g = 2,3). \tag{3.7.42}$$

We assume that that the proper values of the tensor M_{ij} are distinct. We also assume that the amplitudes $a^{(g)}$ are functions of time only. In this case we arrive at the following equations

$$2\rho_0 \frac{\delta_D a^{(g)}}{\delta t} n_i^{(g)} - \frac{1}{U_g^3} b_i^{(g)} (a^{(g)})^2 = (M_{ij} - \rho_0 U_g^2 \delta_{ij}) c_j \tag{3.7.43}$$

$$(g = 2,3),$$

where

$$b_i^{(h)} = B_{MiKjRp}(L)N_M^{(1)}N_K^{(1)}N_R^{(1)}n_j^{(h)}n_p^{(h)} \qquad (h = 2,3). \tag{3.7.44}$$

Analogous to the derivation of (3.7.40), we can show that the amplitude $a^{(h)}$ of each of the transverse waves has the following explicit dependence of time

$$a^{(h)}(t) = a^{(h)}(0)\left(1 - \frac{a^{(h)}(0)b_i^{(h)}n_i^{(h)}}{2\rho_0 U_h^3} t\right)^{-1} \quad (h = 2,3). \tag{3.7.45}$$

If $b_i^{(h)}n_i^{(h)} < 0$, then the amplitude of each of the transverse waves will decay to zero monotonically in infinite time. If $b_i^{(h)}n_i^{(h)} > 0$, then the amplitude of each transverse waves will become infinite within a finite time $t_*^{(h)}$, given by

$$t_*^{(h)} = 2\rho_0 U_h^3 / b_i^{(h)}n_i^{(h)}a^{(h)}(0) \quad (h = 2,3).$$

Clearly, the criteria governing the growth and decay of the amplitude of a transverse wave depend only on the second-order elasticity tensor.

δ) Isotropic materials. In what follows we discuss the propagation of acceleration waves in a homogeneously deformed isotropic elastic solid at rest. The results presented here are due to Green [143].

The response functions α_0, α_1 and α_2 which appear in the constitutive relations (1.3.10) are considered here to depend on the principal invariants I_i^* of the tensor b_{rs}. In the present context the equations of motion are written in the form (1.1.17). The equations (1.3.10) and (1.1.17) imply that the amplitude vector $\underline{a}$ and the local speed of propagation U of an acceleration wave with unit normal $\underline{n}$ obey the following propagation condition

$$(Q_{ij}^*(\underline{n}) - \rho U\delta_{ij})a_j = 0, \tag{3.7.46}$$

where

$$Q_{ij}^*(\underline{n}) = \frac{1}{J} A_{MiKj} x_{r,M} x_{s;K} n_r n_s, \tag{3.7.47}$$

or

$$Q^*_{ij}(\underline{n}) = \frac{1}{J} C_n^2 Q_{ij}(\underline{N}). \tag{3.7.48}$$

We now give the explicit form of (3.7.47) for isotropic solids. By (1.1.19) and (3.7.21),

$$A_{MiKj} = \frac{\partial T_{Mi}}{\partial x_{j,K}} = \frac{\partial (J X_{M,s} t_{si})}{\partial x_{j,K}} .$$

In view of the relations

$$\frac{\partial X_{M,s}}{\partial x_{j,K}} = - X_{M,j} X_{K,s}, \quad \frac{\partial J}{\partial x_{j,K}} = J X_{K,j},$$

we obtain

$$A_{MiKj} = J[(X_{M,s} X_{K,j} - X_{M,j} X_{K,s}) t_{si} + 2F_{isjm} x_{m,K} X_{M,s}], \tag{3.7.49}$$

where

$$\begin{aligned} F_{isjm} = \frac{\partial t_{is}}{\partial b_{jm}} &= \frac{1}{2} \alpha_1 (\delta_{ij}\delta_{sm} + \delta_{sj}\delta_{im}) + \\ &+ \frac{1}{2} \alpha_2 (\delta_{ij} b_{sm} + \delta_{im} b_{sj} + \delta_{sj} b_{im} + \delta_{sm} b_{ij}) + \\ &+ \delta_{is} f^{(o)}_{jm} + b_{is} f^{(1)}_{jm} + b_{ip} b_{ps} f^{(2)}_{jm}, \end{aligned} \tag{3.7.50}$$

$$f^{(h)}_{jm} = \frac{\partial \alpha_h}{\partial I^*_1} \delta_{jm} + \frac{\partial \alpha_h}{\partial I^*_2} (I^*_1 \delta_{jm} - b_{jm}) + I^*_3 \frac{\partial \alpha_h}{\partial I^*_3} b_{jm}$$

$$(h = 0,1,2).$$

It follows from (3.7.47) and (3.7.49) that

$$Q^*_{ij}(\underline{n}) = 2F_{isjm} b_{mr} n_s n_r. \tag{3.7.51}$$

This result is due to Truesdell [355]. In view of (3.7.50) we obtain

$$Q^*_{ij}(\underline{n}) = (\alpha_1 b_{sr} n_s n_r + \alpha_2 b_{sm} b_{mr} n_s n_r)\delta_{ij} + \alpha_1 b_{ir} n_r n_j +$$

$$+ \alpha_2(b_{ir}b_{js}n_rn_s + b_{im}b_{mr}n_rn_j + b_{ij}b_{sr}n_sn_r) +$$

$$+ 2(f^{(o)}_{jm}n_ib_{mr}n_r + f^{(1)}_{jm}b_{is}b_{mr}n_sn_r + f^{(2)}_{jm}b_{ip}b_{ps}b_{mr}n_sn_r).$$

Let $\underline{n}^{(i)}$ (i = 1,2,3) be the principal directions of the deformation tensor b_{ij},

$$b_{ij}n_j^{(p)} = \lambda_p^2 n_i^{(p)} \qquad (p = 1,2,3), \tag{3.7.52}$$

where λ_p is the principal stretch in the pth principal direction. Assume that at every point of the wave surface the propagation direction is parallel to one of the principal directions, say $\underline{n}^{(1)}$, at the spatial point of the body coinciding instantaneously with the surface point. Then, from (3.7.51), (3.7.52) and

$$b_{im}b_{mj}n_j^{(p)} = \lambda_p^4 n_i^{(p)} \qquad (p = 1,2,3),$$

it follows that

$$\begin{aligned} \lambda_1^{-2}Q^*_{ij}(\underline{n}^{(1)}) &= (\alpha_1 + \lambda_1^2\alpha_2)\delta_{ij} + \alpha_2 b_{ij} + \\ &+ 2(\tfrac{1}{2}\alpha_1 + \lambda_1^2\alpha_2 + \sum_{h=0}^{2}\lambda_1^{2h}G_h)n_i^{(1)}n_j^{(1)}, \end{aligned} \tag{3.7.53}$$

where

$$G_p = \frac{\partial\alpha_p}{\partial I_1^*} + (\lambda_2^2 + \lambda_3^2)\frac{\partial\alpha_p}{\partial I_2^*} + \lambda_2^2\lambda_3^2\frac{\partial\alpha_p}{\partial I_3^*}, \quad (p = 0,1,2). \tag{3.7.54}$$

Clearly, the vector characterized by the components $Q^*_{ij}(\underline{n}^{(1)})n_j^{(p)}$ has the direction of the principal direction $\underline{n}^{(p)}$ (p = 1,2,3). Thus, for a wave travelling down a principal axis of strain in an isotropic elastic body, the proper vectors of the acoustic tensor coincide with the principal axes of strain. Since the amplitude vector $\underline{a}$ should be in the direction of one of the proper vectors of the acoustic tensor, it will be in the direction of one of $\underline{n}^{(1)}$, $\underline{n}^{(2)}$ or $\underline{n}^{(3)}$. If $\underline{a}$ is parallel to $\underline{n}^{(1)}$, the direction of propagation, then the acceleration wave is said to be longitudinal. If $\underline{a}$

is parallel to either $\underline{n}^{(2)}$ or $\underline{n}^{(3)}$ it is perpendicular to $\underline{n}^{(1)}$ and is said to be transverse. Following the previous terminology we call such a wave a principal wave. We can state that every principal wave is either longitudinal or transverse. For the $\underline{n}^{(1)}$ principal direction, the speed U_{11} of the longitudinal wave is given by

$$\rho U_{11}^2 = Q_{ij}^*(\underline{n}^{(1)})n_i^{(1)}n_j^{(1)} = 2\lambda_1^2(\alpha_1 + 2\lambda_1^2\alpha_2 + \sum_{h=0}^{2}\lambda_1^{2h}G_h). \qquad (3.7.55)$$

The speed U_{12} of the transverse wave with amplitude vector in the $\underline{n}^{(2)}$ principal direction is given by

$$\rho U_{12}^2 = Q_{ij}^*(\underline{n}^{(1)})n_i^{(2)}n_j^{(2)} = \lambda_1^2[\alpha_1 + (\lambda_1^2 + \lambda_2^2)\alpha_2]. \qquad (3.7.56)$$

Similarly, for the speed U_{13} we obtain

$$\rho U_{13}^2 = \lambda_1^2[\alpha_1 + (\lambda_1^2 + \lambda_3^2)\alpha_2].$$

The principal stresses are given by

$$t_i = \alpha_o + \lambda_i^2\alpha_1 + \lambda_i^4\alpha_2.$$

By using the relations

$$I_1^* = \lambda_1^2 + \lambda_2^2 + \lambda_3^2, \quad I_2^* = \lambda_1^2\,\lambda_2^2 + \lambda_1^2\,\lambda_3^2 + \lambda_2^2\,\lambda_3^2, \; I_3^* = \lambda_1^2\,\lambda_2^2\,\lambda_3^2,$$

we obtain

$$\frac{\partial t_1}{\partial \lambda_1} = 2\lambda_1\,\alpha_1 + 4\lambda_1^3\alpha_2 + \sum_{p=0}^{2}\lambda_1^{2p}\frac{\partial \alpha_p}{\partial \lambda_1} = 2\lambda_1(\alpha_1 + 2\lambda_1^2\,\alpha_2 + \sum_{p=0}^{2}\lambda_1^{2p}G_p).$$

Thus

$$\rho U_{11} = \lambda_1\frac{\partial t_1}{\partial \lambda_1}\,.$$

In terms of the differences t_1-t_2 and t_1-t_3, we have

$$\rho U_{12}^2 = \lambda_1^2 \frac{t_1-t_2}{\lambda_1^2 - \lambda_2^2}, \qquad \rho U_{13}^2 = \lambda_1^2 \frac{t_1-t_3}{\lambda_1^2 - \lambda_3^2}.$$

In order to study the behaviour of the amplitude a we shall need explicit expression for the second-order elasticity tensor. Analogous to the derivation of (3.7.49), we can show that

$$\begin{aligned} B_{MiKjLr} = J\{[&X_{L,r}(X_{M,s}X_{K,j} - X_{M,j}X_{K,s}) - X_{L,s}(X_{M,r}X_{K,j} - \\ &- X_{M,j}X_{K,r}) - X_{L,j}(X_{M,s}X_{K,r} - X_{M,r}X_{K,s})]t_{si} + \\ &+ 2x_{p,L}(X_{M,n}X_{K,j} - X_{M,j}X_{K,n})F_{inrp} + \\ &+ 2x_{p,K}(X_{M,n}X_{L,r} - X_{M,r}X_{L,n})F_{injp} + \\ &+ 2\delta_{KL}X_{M,n}F_{injr} + 4X_{M,n}x_{p,K}x_{s,L}F_{injprs}\}, \end{aligned} \tag{3.7.57}$$

where

$$F_{krmnpq} = \frac{\partial F_{krmn}}{\partial b_{pq}}.$$

It follows from (3.7.50) that

$$\begin{aligned} F_{ksmnpq} = \frac{1}{4}\,\alpha_2[&(\delta_{kp}\delta_{qr} + \delta_{kq}\delta_{pr})(\delta_{rm}\delta_{sn} + \delta_{rn}\delta_{sm}) + \\ &+ (\delta_{sq}\delta_{pr} + \delta_{sp}\delta_{qr})(\delta_{nr}\delta_{km} + \delta_{mr}\delta_{kn})] + \\ &+ \delta_{ks}F^{(o)}_{mnpq} + b_{ks}F^{(1)}_{mnpq} + b_{ki}b_{is}F^{(2)}_{mnpq} + \\ &+ \tfrac{1}{2}(\delta_{kp}\delta_{sq} + \delta_{kq}\delta_{sp})H^{(1)}_{mn} + \tfrac{1}{2}(\delta_{km}\delta_{sn} + \delta_{kn}\delta_{sm})H^{(1)}_{pq} + \\ &+ \tfrac{1}{2}(\delta_{sq}b_{kp} + \delta_{sp}b_{kq} + \delta_{kq}b_{sp} + \delta_{kp}b_{sq})H^{(2)}_{mn} + \end{aligned} \tag{3.7.58}$$

$$+ \frac{1}{2}(\delta_{sn}b_{km} + \delta_{sm}b_{kn} + \delta_{kn}b_{sm} + \delta_{km}b_{sn})H^{(2)}_{pq},$$

where

$$H^{(g)}_{mn} = P_{g1}\delta_{mn} + P_{g2}D_{mn} + I^*_3 P_{g3} b^{(-1)}_{mn},$$

$$F^{(g)}_{ksmn} = P_{g2}[\delta_{ks}\delta_{mn} - \frac{1}{2}(\delta_{km}\delta_{sn} + \delta_{kn}\delta_{sm})] +$$

$$+ P_{g3}[I^*_1\delta_{ks}\delta_{mn} - \delta_{ks}b_{mn} - \delta_{mn}b_{ks} - \frac{1}{2} I^*_1(\delta_{km}\delta_{sn} +$$

$$+ \delta_{kn}\delta_{sm}) + \frac{1}{2}(\delta_{km}b_{sn} + \delta_{kn}b_{sm} + \delta_{sn}b_{km} + \delta_{sm}b_{kn})] +$$

$$+ (R_{g11}\delta_{mn} + R_{g12}D_{mn} + I^*_3 R_{g13} b^{(-1)}_{mn})]\delta_{ks} +$$

$$+ (R_{g12}\delta_{mn} + R_{g22}D_{mn} + I^*_3 R_{g23} b^{(-1)}_{mn})D_{ks} +$$

$$+ (R_{g13}\delta_{mn} + R_{g23}D_{mn} + I^*_3 R_{g33} b^{(-1)}_{mn}) I^*_3 b^{(-1)}_{ks},$$

$$P_{gi} = \frac{\partial\alpha_g}{\partial I^*_i}, \quad R_{gij} = \frac{\partial^2\alpha_g}{\partial I^*_i\, \partial I^*_j}, \quad (g = 0,1,2), \quad D_{mn} = I^*_1\, \delta_{mn} - b_{mn}.$$

We shall study the properties of principal waves propagating in the $\underline{n}^{(1)}$ principal direction. We assume that the waves propagate in a homogeneously deformed, isotropic elastic solid at rest. Thus, we suppose that the stretches λ_i are constants. Moreover, we assume that the discontinuity across each of the waves is uniform; that is we consider that the magnitude of the discontinuity is independent of the position in the wave plane.

It follows from (3.7.37), (3.7.39), (3.7.40), (3.7.57) and (3.7.58) that the amplitude a of the longitudinal wave propagating in the $\underline{n}^{(1)}$ direction is given by

$$a(t) = a(0)[1 - a(0)qt]^{-1},$$

where

$$q = \frac{1}{2\rho U_{11}^3}\left(\rho U_{11}^2 + \frac{\partial^2 t_1}{\partial \lambda_1^2} - \frac{1}{\lambda_1}\frac{\partial t_1}{\partial \lambda_1}\right).$$

We can see that the transverse principal waves propagate without any change in their amplitude.

In Section 3.5 we have studied the propagation of plane waves of infinitesimal amplitude in a homogeneously deformed elastic body. By using the relations (1.1.15), (1.1.19), (1.2.30) and (1.2.44), it may be shown that the propagation condition (3.5.3) is nothing more than the equation (3.7.46). Thus, the propagation condition for plane waves of infinitesimal amplitude in a homogeneously deformed body is the same as those of acceleration waves. This result is due to Truesdell [355]. The amplitude of a plane infinitesimal wave propagating in any direction in a homogeneously deformed elastic body remains unchanged as the wave travels in the material. For plane acceleration waves this result is only valid if γ, as defined by (3.7.39), vanishes for all directions of propagation.

In [371], Wesolowski has studied the vibrations of an isotropic elastic medium deforming in time; the body B_0 is subjected to finite strains in such a manner that the elongations in three mutually perpendicular directions are proportional to time. It is shown that three principal directions of propagation exist connected with longitudinal and transversal waves. Other results on propagation of acceleration waves in homogeneously deformed elastic materials are given by Green [138], Chen [76], Eringen and Şuhubi [116], McCarthy [257] and Wesolowski [375].

4 Thermoelastic materials

4.1 Thermodynamic preliminaries

We now study the thermo-mechanical behaviour of the body. We postulate an energy balance in the form (see Coleman and Noll [82])

$$\int_P \rho_0(\dot{x}_i\ddot{x}_i + \dot{\varepsilon})dV = \int_P \rho_0 f_i\dot{x}_i dV + \int_{\partial P} T_i\dot{x}_i dA + \int_P \rho_0\, s dV + \int_{\partial P} HdA, \tag{4.1.1}$$

for every part P of B_0 and every time t. Here we have introduced the following additional quantities associated with the motion of the body: H is the heat flux across the surface ∂P, measured per unit area of ∂P, and s is the heat supply per unit mass and unit time. With an argument similar to that used in Section 1.1.1 we obtain the relations (1.1.5) and the equations of motion (1.1.6). In view of (1.1.5) and (1.1.6) the relation (4.1.1) reduces to

$$\int_P \rho_0\, \dot{\varepsilon}\, dV = \int_P (\rho_0 s + T_{Ai}\dot{x}_{i,A})dV + \int_{\partial P} HdA. \tag{4.1.2}$$

We apply this equation to a region P which is such that in P it was a tetrahedron element bounded by a plane with the unit normal N_A, and by planes through the point $\underline{X}$ parallel to the coordinate planes. We find that

$$H = Q_A N_A, \tag{4.1.3}$$

where Q_A is the heat flux across surfaces that were originally coordinate planes perpendicular to the X_A-axes, measured per unit area of these planes per unit time. Use of this relation in (4.1.2) then leads to the local equation of energy

$$\rho_0\dot{\varepsilon} = T_{Ai}\dot{x}_{i,A} + \rho_0 s + Q_{A,A}. \tag{4.1.4}$$

The relation (1.1.10) follows from (4.1.4) with an argument similar to that used in obtaining (1.1.10) from (1.1.9). With the help of (1.1.10) and (1.1.11), the energy equation (4.1.4) reduces to

$$\rho_0 \dot{\varepsilon} = T_{AB}\dot{E}_{AB} + \rho_0 s + Q_{A,A}. \tag{4.1.5}$$

The second law of thermodynamics is the assertion that (see, for example, Green and Zerna [140], Sect. 2.8, Truesdell and Noll [356], Sect. 79)

$$\int_P \rho_0 \dot{\eta}\, dV - \int_P \frac{1}{T} \rho_0\, s dV - \int_{\partial P} \frac{1}{T} H dA \geqq 0, \tag{4.1.6}$$

for every part P of B_0 and every time t. Here η is the entropy per unit mass, and T is the absolute temperature, which is assumed to be always positive. Using (4.1.3) in (4.1.6), transforming the surface integral to a volume integral, and making the usual smoothness assumption, we obtain the local entropy inequality

$$\rho_0 T\dot{\eta} - \rho_0 s - Q_{A,A} + \frac{1}{T} Q_A T_{,A} \geqq 0. \tag{4.1.7}$$

Introducing the Helmholtz free-energy per unit of initial volume,

$$\psi = \rho_0(\varepsilon - T\eta), \tag{4.1.8}$$

and combining (4.1.5) and (4.1.7), we arrive at the local dissipation inequality

$$T_{AB}\dot{E}_{AB} - \rho_0 \eta \dot{T} - \dot{\psi} + \frac{1}{T} Q_A T_{,A} \geqq 0. \tag{4.1.9}$$

A thermoelastic material is defined as one for which the following constitutive equations hold at each point $\underline{X}$ and for all time t

$$\begin{aligned} \psi &= \tilde{\psi}(x_{i,A}, T, T_{,B}), \qquad T_{MN} = \tilde{T}_{MN}(x_{i,A}, T, T_{,B}), \\ \eta &= \tilde{\eta}(x_{i,A}, T, T_{,B}), \qquad Q_M = \tilde{Q}_M(x_{i,A}, T, T_{,B}). \end{aligned} \tag{4.1.10}$$

We assume that the response functions are of class C^2 on their domain, which is the set of all $(x_{i,A},T,T_{,B})$ with $\det(x_{i,A}) \neq 0$, $T > 0$, $\underline{X} \in B_0$. The constitutive equations must satisfy invariance conditions under superposed rigid-body motions. Hence

$$\tilde{\psi}(x_{i,A},T,T_{,B}) = \tilde{\psi}(Q_{ir}x_{r,A},T,T_{,B}), \tag{4.1.11}$$

for all proper orthogonal tensors Q_{ir}. The functions $\tilde{T}_{AB}$, $\tilde{\eta}$ and $\tilde{Q}_K$ must satisfy the same invariance condition as $\tilde{\psi}$. In view of (1.1.2) we may write $x_{i,A}$ in the polar form

$$x_{i,A} = R_{iB}M_{BA},$$

where M_{BA} is a positive definite symmetric tensor and R_{iB} is a rotation tensor, so that

$$R_{iB}R_{iA} = \delta_{AB}, \quad R_{iA}R_{jA} = \delta_{ij}, \quad \det (R_{iA}) = 1.$$

We may take the special value $R_{jK}\delta_{Kr}$ for Q_{rj} in (4.1.11) so that

$$\tilde{\psi}(x_{i,A},T,T_{,B}) = \tilde{\psi}(M_{KL},T,T_{,B}).$$

Recalling (1.1.11),

$$2E_{AB} = M_{AK}M_{KB} - \delta_{AB}.$$

Since M_{AB} is a positive definite symmetric tensor, a single valued function of M_{AB} can be replaced by a single-valued function of E_{AB}, so that we may replace ψ by a different function

$$\psi = \hat{\psi}(E_{AB},T,T_{,K}). \tag{4.1.12}$$

We can verify that this satisfies the condition (4.1.11) for arbitrary proper orthogonal values of Q_{ij}.

With the help of (4.1.12), the inequality (4.1.9) becomes

$$(T_{AB} - \frac{\partial\psi}{\partial E_{AB}})\dot{E}_{AB} - (\rho_0\eta + \frac{\partial\psi}{\partial T})\dot{T} - \frac{\partial\psi}{\partial T_{,B}}\dot{T}_{,B} + \frac{1}{T}Q_A T_{,A} \geqq 0. \qquad (4.1.13)$$

To avoid ambiguity we assume that ψ in (4.1.13) is arranged as a symmetric function of E_{AB}. For a given deformation and temperature, the inequality (4.1.13) is valid for all arbitrary values of $\dot{T}, \dot{T}_{,A}$ and $\dot{E}_{AB}$, subject to $E_{AB} = E_{BA}$. Then (see, for example, Carlson [47]),

$$T_{AB} - \frac{\partial\psi}{\partial E_{AB}} = 0, \quad \rho_0\eta + \frac{\partial\psi}{\partial T} = 0, \quad \frac{\partial\psi}{\partial T_{,B}} = 0,$$

and

$$Q_A T_{,A} \geqq 0. \qquad (4.1.14)$$

We conclude that the constitutive equations of thermoelastic bodies are of the form

$$\psi = \hat{\psi}(E_{AB}, T), \quad T_{AB} = \frac{\partial\psi}{\partial E_{AB}}, \quad \rho_0\eta = -\frac{\partial\psi}{\partial T},$$
$$Q_A = \hat{Q}_A(E_{AB}, T, T_{,M}). \qquad (4.1.15)$$

The restriction placed on the response functions by the second law have been studied by Green and Adkins [135], Coleman and Noll [82], Coleman and Mizel [83] see, also, Carlson [47]). With the help of (4.1.8) and (4.1.15), the energy equation (4.1.5) reduces to

$$\rho_0 T\dot{\eta} = Q_{A,A} + \rho_0 S, \qquad (4.1.16)$$

on $B_0 \times (t_0, t_1)$. The heat conduction inequality (4.1.14) implies that

$$\hat{Q}_M(E_{AB}, T, 0) = 0. \qquad (4.1.17)$$

Pipkin and Rivlin [292] who first obtained (4.1.17), described it as showing the "non-existence of a piezocaloric effect".

The complete system of field equations for nonlinear thermoelasticity consists of the equations of motion (1.1.6), the energy equation (4.1.16), the constitutive equations (4.1.15), and the geometrical equations (1.1.11).

Of course, the response functions $\hat{Q}_A$ are subject to the restriction (4.1.14).

We consider the following boundary conditions

$$x_i = \hat{x}_i \text{ on } \bar{\Sigma}_1 \times [t_o,t_1), \quad T_{Ai}N_A = \hat{T}_i \text{ on } \Sigma_2 \times [t_o,t_1),$$
$$T = \hat{T} \text{ on } \bar{\Sigma}_3 \times [t_o,t_1), \quad Q_AN_A = \hat{Q} \quad \text{on } \Sigma_4 \times [t_o,t_1), \tag{4.1.18}$$

where Σ_3 and Σ_4 are complementary subsurfaces of ∂B_o as are Σ_1 and Σ_2, and $\hat{x}_i, \hat{T}_i, \hat{T}$ and $\hat{Q}$ are prescribed functions.

It is sometimes useful to have the above relations expressed in terms of quantities referred to the present configuration B. These quantities are the Cauchy stress tensor and the heat flux q_i across the x_i-plane at $\underline{x}$, per unit area of this plane and per unit time. We record the relation (see, for example, Green and Adkins [135], Sect. 8.1, Eringen [117], Sect. 2.4)

$$Jq_i = x_{i,A}Q_A. \tag{4.1.19}$$

The equations of motions (1.1.6) are now replaced by (1.1.17), and the energy equation (4.1.16) becomes

$$\rho T\dot{\eta} = q_{i,i} + \rho s \text{ on } B \times (t_o,t_1). \tag{4.1.20}$$

The boundary conditions corresponding to (4.1.18) are

$$x_i = \tilde{x}_i \text{ on } \bar{S}_1 \times [t_o,t_1), \quad t_{ji}n_j = \tilde{t}_i \text{ on } S_2 \times [t_o,t_1),$$
$$T = \tilde{T} \quad \text{on } \bar{S}_3 \times [t_o,t_1), \quad q_in_i = \tilde{q} \quad \text{on } S_4 \times [t_o,t_1), \tag{4.1.21}$$

where $\tilde{x}_i, \tilde{t}_i, \tilde{T}$ and q are prescribed functions, and S_r is the image of Σ_r $(r = 1,2,3,4)$ by motion.

4.2 Infinitesimal thermoelastic deformations superimposed upon a given deformation

In this section we derive the equations governing infinitesimal thermoelastic deformations superimposed on large deformations. The theory of infinitesimal thermoelastic deformations in a body which has already been subjected to an isothermal elastic deformation was established by England and Green [111] and Green [137]. The equations established in [111], [137], have been employed by Knops and Wilkes [229] in a discussion of thermoelastic stability. The theory of infinitesimal thermoelastic deformations superimposed on a large deformation at nonconstant temperature was established by Ieşan [195].

We consider three states of the body: the state B_0, the primary state B, and the secondary state B* corresponding respectively to constant temperature T_0, temperature T, and temperature T*. We assume that B_0 is an unstressed equilibrium configuration. As in Section 1.2, the thermodynamic quantities associated with the secondary state will be denoted with an asterisk. The position coordinates of the particle $\underset{\sim}{X}$ at time t in B* will be denoted by y_i. Now

$$u_i = y_i - x_i, \quad \theta = T^* - T. \tag{4.2.1}$$

We assume that u_i and θ are small, i.e.

$$u_i = \varepsilon' u_i', \quad \theta = \varepsilon' \theta', \tag{4.2.2}$$

where ε' is a constant small enough for squares and higher powers to be neglected, and u_i' and θ' are independent of ε'.

The problem consists in establishing the equations, boundary conditions, and initial conditions for u_i and θ when the primary deformation and the loadings associated with B and B* are known.

In the secondary state B* we consider the following stress tensors: t_{ij}^* is the Cauchy stress tensor; T_{Ai}^* and T_{AB}^* are the Piola-Kirchhoff stress tensors measured per unit area in the configuration B_0; $T_{ki}^{*(1)}$ is the first Piola-Kirchhoff stress tensor measured per unit area in the configuration B. We also consider the following heat flux vectors: Q_A^* is the heat flux across surfaces in B* that in the configuration B_0 were coordinate planes perpendicular to the X_A axes, measured per unit area of these planes and per unit time; $\bar{Q}_i^*$ is the heat flux across surfaces in B* that in the configuration B were coordinate planes perpendicular to the x_i axes, measured per unit area of these planes and per unit time. Then,

$$q_i^* = \frac{1}{J^*} y_{i,A} Q_A^* = \frac{1}{J'} y_{i,j} \bar{Q}_j^*, \tag{4.2.3}$$

where J* and J' are given by (1.2.8).

The equations of motion for the secondary state may be written in the forms (1.2.10) and (1.2.11). The energy equation governing the secondary state can be written in the forms

$$\rho_o T^* \dot{\eta}^* = Q^*_{A,A} + \rho_o s^* \quad \text{on } B_o \times (t_o, t_1), \tag{4.2.4}$$

$$\rho T^* \dot{\eta}^* = \bar{Q}^*_{i,i} + \rho s^* \quad \text{on } B \times (t_o, t_1). \tag{4.2.5}$$

The constitutive equations imply that

$$\begin{aligned} \psi^* &= \hat{\psi}(E^*_{AB}, T^*), \quad T^*_{AB} = \frac{\partial \psi^*}{\partial E^*_{AB}}, \\ \rho_o \eta^* &= -\frac{\partial \psi^*}{\partial T^*}, \quad Q^*_A = \hat{Q}_A(E^*_{AB}, T^*, T^*_{,A}), \end{aligned} \tag{4.2.6}$$

where E^*_{AB} are given by (1.2.13).

The heat conduction inequality leads to

$$q_i T_{,i} \geqq 0, \quad Q_A T_{,A} \geqq 0, \quad Q^*_A T^*_{,A} \geqq 0. \tag{4.2.7}$$

We consider the following boundary conditions for the secondary deformation

$$\begin{aligned} y_i &= \hat{y}_i \text{ on } \bar{\Sigma}_1 \times [t_o, t_1), \quad T^*_{Ai} N_A = \hat{T}^*_i \text{ on } \Sigma_2 \times [t_o, t_1), \\ T^* &= \hat{T}^* \text{ on } \bar{\Sigma}_3 \times [t_o, t_1), \quad Q^*_A N_A = \hat{Q}^* \text{ on } \Sigma_4 \times [t_o, t_1), \end{aligned} \tag{4.2.8}$$

where $\hat{y}_i$, $\hat{T}^*_i$, $\hat{T}^*$ and $\hat{Q}^*$ are prescribed functions. The conditions (4.2.8) can be written in the form

$$\begin{aligned} y_i &= \tilde{y}_i \text{ on } \bar{S}_1 \times [t_o, t_1), \quad T^{*(1)}_{ji} n_j = \tilde{T}^*_i \text{ on } S_2 \times [t_o, t_1), \\ T^* &= \tilde{T}^* \text{ on } \bar{S}_3 \times [t_o, t_1), \quad \bar{Q}^*_i n_i = \tilde{q}^* \quad \text{on } S_4 \times [t_o, t_1), \end{aligned} \tag{4.2.9}$$

where $\tilde{y}_i$, $\tilde{T}^*_i$, $\tilde{T}^*$ and $\tilde{q}^*$ are specified functions.

If we use (4.2.2) and (1.2.16), then to a second order approximation, we obtain

$$\frac{\partial\psi^*}{\partial E^*_{KL}} = \frac{\partial\psi}{\partial E_{KL}} + A_{KLMN}(E^*_{MN} - E_{MN}) - B_{KL}\theta,$$
$$\frac{\partial\psi^*}{\partial T^*} = \frac{\partial\psi}{\partial T} - B_{MN}(E^*_{MN} - E_{MN}) - A\theta, \tag{4.2.10}$$

where

$$A_{KLMN} = \frac{\partial^2\psi}{\partial E_{KL}\,\partial E_{MN}}, \quad B_{KL} = -\frac{\partial^2\psi}{\partial E_{KL}\,\partial T}, \quad A = -\frac{\partial^2\psi}{\partial T^2}. \tag{4.2.11}$$

Clearly,

$$A_{KLMN} = A_{MNKL} = A_{LKMN}, \quad B_{KL} = B_{LK}. \tag{4.2.12}$$

From (4.1.15), (4.2.6), (1.2.16) or (1.2.17), and (4.2.10) we obtain

$$T^*_{KL} = T_{KL} + A_{KLMN}x_{i,M}u_{i,N} - B_{KL}\theta,$$
$$\rho_0\eta^* = \rho_0\eta + B_{MN}x_{i,M}u_{i,N} + A\theta, \tag{4.2.13}$$

or

$$T^*_{KL} = T_{KL} + A_{KLMN}x_{i,M}x_{j,N}e_{ij} - B_{KL}\theta,$$
$$\rho_0\eta^* = \rho_0\eta + B_{MN}x_{i,M}x_{j,N}e_{ij} + A\theta. \tag{4.2.14}$$

It follows from (1.1.8) and (4.2.13) that

$$T^*_{Ki} = T_{Ki} + E_{iKjN}u_{j,N} - F_{Ki}\theta + T_{KN}u_{i,N}, \tag{4.2.15}$$

where

$$E_{iKjN} = x_{i,B}x_{j,M}A_{BKMN}, \quad F_{Ki} = B_{KL}x_{i,L}. \tag{4.2.16}$$

By (1.2.26) and (4.2.15),

$$S_{Ai} = E_{iAjN}u_{j,N} - F_{Ai}\theta + T_{KN}u_{i,N}. \tag{4.2.17}$$

From (1.1.19), (1.2.28) and (4.2.14) we obtain

$$T^{*(1)}_{ij} = t_{ij} + C_{ijrs}e_{rs} - \beta_{ij}\theta + t_{ir}u_{j,r}, \tag{4.2.18}$$

where

$$JC_{ijrs} = x_{i,K}x_{j,L}x_{r,M}x_{s,N}A_{KLMN}, \quad J\beta_{ij} = x_{i,K}x_{j,L}B_{KL}. \tag{4.2.19}$$

Obviously, C_{ijrs} and β_{ij} are symmetric tensors. It follows from (1.2.32) and (4.2.18) that

$$s_{ij} = C_{ijrs}u_{r,s} - \beta_{ij}\theta + t_{ir}u_{j,r}. \tag{4.2.20}$$

Next, define

$$\gamma = \rho_0(\eta^*-\eta), \; \Phi_A = Q^*_A - Q_A, \; \phi_i = \overline{Q}^*_i - q_i. \tag{4.2.21}$$

By (4.2.13), (4.2.14), (4.2.16) and (4.2.20) we obtain

$$\gamma = F_{Ai}u_{i,A} + A\theta, \tag{4.2.22}$$

$$\rho(\eta^*-\eta) = \frac{1}{J}\gamma = \beta_{ij}e_{ij} + a\theta = \beta_{ij}u_{i,j} + a\theta, \tag{4.2.23}$$

where

$$a = \frac{1}{J}A. \tag{4.2.24}$$

We now consider the heat flux vectors. In view of (4.2.2) and (4.2.6),

$$Q^*_A = Q_A + R_{AMN}(E^*_{MN} - E_{MN}) + D_A\theta + K_{AM}\theta_{,M}, \tag{4.2.25}$$

where

$$R_{AMN} = \frac{\partial Q_A}{\partial E_{MN}} = R_{ANM}, \quad D_A = \frac{\partial Q_A}{\partial T}, \quad K_{AB} = \frac{\partial Q_A}{\partial T_{,B}}. \tag{4.2.26}$$

We note that (4.1.17) implies

$$R_{AMN} = 0, \quad D_A = 0 \quad \text{if} \quad T_{,K} = 0. \tag{4.2.27}$$

From (1.2.16), (4.2.21) and (4.2.25) we find

$$\Phi_A = G_{AiN}u_{i,N} + D_A\theta + K_{AM}\,\theta_{,M}, \tag{4.2.28}$$

where

$$G_{AiN} = R_{AMN}x_{i,M}. \tag{4.2.29}$$

By (1.2.17) and (4.2.25),

$$Q_A^* = Q_A + R_{AMN}x_{i,M}x_{j,N}e_{ij} + D_A\theta + K_{AM}x_{i,M}\theta_{,i}. \tag{4.2.30}$$

The relation (4.2.3) implies that

$$\bar{Q}_i^* = \frac{1}{J}\,x_{i,A}\,Q_A^*. \tag{4.2.31}$$

Using (4.1.19), (4.2.21), (4.2.30) and (4.2.31) we obtain

$$\phi_i = h_{irs}u_{r,s} + a_i\theta + k_{ij}\,\theta_{,j}, \tag{4.2.32}$$

where

$$h_{irs} = \frac{1}{J}\,x_{i,K}x_{r,L}x_{s,M}R_{KLM}, \quad a_i = \frac{1}{J}\,x_{i,K}D_K,$$
$$k_{ij} = \frac{1}{J}\,x_{i,A}x_{j,B}K_{AB}. \tag{4.2.33}$$

It follows from (4.2.27) and (4.2.33) that

$$h_{irs} = 0, \quad a_i = 0 \quad \text{if} \quad T_{,j} = 0. \tag{4.2.34}$$

In view of (4.2.7) and (4.2.31),

$$\bar{Q}_i^* T^*_{,i} \geq 0,$$

so that we have

$$q_i T_i + h_{irs} e_{rs} T_{,i} + a_i \theta\, T_{,i} + k_{ij}\, \theta_{,j} T_{,i} + q_i\, \theta_{,i} + \\ + h_{irs} e_{rs} \theta_{,i} + a_i \theta\, \theta_{,i} + k_{ij}\, \theta_{,i}\, \theta_{,j} \geq 0, \tag{4.2.35}$$

for any small quantities e_{ij} and θ. From (4.2.34) and (4.2.35) we find that

$$k_{ij} \theta_{,i}\, \theta_{,j} \geq 0 \text{ if } T_{,s} = 0. \tag{4.2.36}$$

α) <u>The equations of thermoelastic deformations referred to the configuration B_o</u>. If we refer all quantities to the configuration B_o then we have

$$u_i = u_i(X_A,t),\ \theta = \theta(X_A,t),\quad (X_A,t) \in B_o \times (t_o,t_1). \tag{4.2.37}$$

Forming differences of corresponding terms in the two equations (4.1.16) and (4.2.4), with the aid of (4.2.21), we obtain

$$T\dot{\gamma} + \rho_o \dot{\eta} = \Phi_{A,A} + \rho_o S,\quad \text{on } B_o \times (t_o,t_1), \tag{4.2.38}$$

where

$$S = s^* - s. \tag{4.2.39}$$

From the equations of motion we find the equations (1.2.34). Thus, the basic equations consist of the equations of motion (1.2.34), the energy equation (4.2.38), and the "constitutive" equations (4.2.17), (4.2.22) and (4.2.28). It is important to realize that the coefficients E_{iAjN}, F_{Ai}, A, G_{AiK}, D_A and K_{AM} depend on the state of strain and temperature of the primary state. From (4.1.18) and (4.2.8) we obtain the boundary conditions

$$u_i = \hat{u}_i \text{ on } \bar{\Sigma}_1 \times [t_o,t_1), \quad S_{Ai}N_A = \hat{P}_i \text{ on } \Sigma_2 \times [t_o,t_1),$$
$$\theta = \hat{\theta} \text{ on } \bar{\Sigma}_3 \times [t_o,t_1), \quad \Phi_A N_A = \hat{\Phi} \text{ on } \Sigma_4 \times [t_o,t_1), \tag{4.2.40}$$

where

$$\hat{u}_i = \hat{y}_i - \hat{x}_i, \quad \hat{P}_i = \hat{T}_i^* - \hat{T}_i, \; \hat{\theta} = \hat{T}^* - \hat{T}, \; \hat{\Phi} = \hat{Q}^* - \hat{Q}.$$

To complete the specification of the problem we adjoin the initial conditions

$$u_i(\underline{X},0) = \hat{a}_i(\underline{X}), \quad \dot{u}_i(\underline{X},0) = \hat{b}_i(\underline{X}), \; \theta(\underline{X},0) = \hat{\theta}_o(\underline{X}), \quad \underline{X} \in B_o,$$

where $\hat{a}_i$, $\hat{b}_i$ and $\hat{\theta}_o$ are prescribed functions.

β) <u>The equations of thermoelasticity referred to the primary state.</u> If we refer the thermodynamic quantities to the primary state, then we have

$$u_i = u_i(x_j,t), \; \theta = \theta(x_j,t) \quad (x_j,t) \in B \times (t_o,t_1). \tag{4.2.41}$$

Forming differences of corresponding terms in the two equations (4.1.20) and (4.2.5), with the aid of (4.2.21) and (4.2.39), we obtain

$$\frac{1}{J} T\dot{\gamma} + \rho\dot{\eta}_\theta = \phi_{i,i} + \rho S \quad \text{on } B \times (t_o,t_1). \tag{4.2.42}$$

The equations of motion (1.1.17) and (1.2.11) lead to the equations (1.2.42). In this case, the basic equations consist of the equations of motion (1.2.42), the energy equation (4.2.42), and the "constitutive" equations (4.2.20), (4.2.23) and (4.2.32).

By using (4.1.21) and (4.2.9) we derive the boundary conditions

$$u_i = \tilde{u}_i \text{ on } \bar{S}_1 \times [t_o,t_1), \; s_{ji}n_j = \tilde{p}_i \text{ on } S_2 \times [t_o,t_1),$$
$$\theta = \tilde{\theta} \text{ on } \bar{S}_3 \times [t_o,t_1), \; \phi_i n_i = \tilde{\phi} \text{ on } S_4 \times [t_o,t_1), \tag{4.2.43}$$

where

$$\tilde{u}_i = \tilde{y}_i - \tilde{x}_i, \quad \tilde{p}_i = \tilde{T}^*_i - \tilde{T}_i, \quad \tilde{\theta} = \tilde{T}^* - \tilde{T}, \quad \tilde{\phi} = \tilde{q}^* - \tilde{q}.$$

To complete the specification of the problem given by (1.2.42), (4.2.42), (4.2.20), (4.2.23), (4.2.32) and (4.2.43) we must adjoin the initial conditions

$$u_i(\underline{x},0) = \tilde{a}_i(\underline{x}), \quad \dot{u}_i(\underline{x},0) = \tilde{b}_i(\underline{x}), \quad \theta(\underline{x},0) = \tilde{\theta}_0(\underline{x}), \qquad \underline{x} \in B, \tag{4.2.44}$$

where $\tilde{a}_i$, $\tilde{b}_i$ and $\tilde{\theta}_0$ are specified functions.

To close this section we add an expression for the free energy function ψ^* corresponding to the secondary state B^*. Since ψ^* is a function of $T + \theta$ and the strain tensor E^*_{AB} given by (1.2.48), retaining only linear and quadratic terms, we find that

$$\psi^* = \psi + J[t_{ij}e_{ij} - \rho\eta\theta + \tfrac{1}{2}(C_{ijrs}e_{ij}e_{rs} - 2\beta_{ij}e_{ij}\theta - a\theta^2 + t_{ij}u_{r,i}u_{r,j})]. \tag{4.2.45}$$

The function ψ^* may be written in the form

$$\psi^* = \psi + T_{Ai}u_{i,A} - \rho_0\eta\theta + \tfrac{1}{2}(E_{iKjN}u_{i,K}u_{j,N} - 2F_{Ai}u_{i,A}\theta - A\theta^2 + T_{AB}u_{i,A}u_{i,B}).$$

4.3 Isotropic materials

When the body is initially isotropic, the free energy function has the form

$$\psi = \hat{\psi}(I_1, I_2, I_3, T), \tag{4.3.1}$$

where I_s are given by (1.3.2). It follows from (4.1.15) and (4.3.1) that

$$T_{AB} = \hat{\gamma}_{-1}C^{(-1)}_{AB} + \hat{\gamma}_o\,\delta_{AB} + \hat{\gamma}_1 C_{AB}, \tag{4.3.2}$$

where

$$\hat{\gamma}_{-1} = 2I_3 \frac{\partial\hat{\psi}}{\partial I_3}, \quad \hat{\gamma}_o = 2\left(\frac{\partial\hat{\psi}}{\partial I_1} + I_1 \frac{\partial\hat{\psi}}{\partial I_2}\right), \quad \hat{\gamma}_1 = -2 \frac{\partial\hat{\psi}}{\partial I_2}.$$

The derivation is similar to that in the Section 1.3. Clearly,

$$t_{ij} = \hat{\Phi} b_{ij} + \hat{\Psi} B_{ij} + \hat{p}\delta_{ij}, \tag{4.3.3}$$

where

$$\hat{\Phi} = \frac{2}{\sqrt{I_3}} \frac{\partial\hat{\Psi}}{I_1}, \quad \hat{\Psi} = \frac{2}{\sqrt{I_3}} \frac{\partial\hat{\Psi}}{\partial I_2}, \quad \hat{p} = 2\sqrt{I_3} \frac{\partial\hat{\Psi}}{\partial I_3}. \tag{4.3.4}$$

From (1.3.14), (4.1.15), (4.2.11) and (4.3.2) we obtain

$$\begin{aligned}
A_{KLMN} &= \bar{A}^{(-1)}_{MN} C^{(-1)}_{KL} + \bar{A}^{(0)}_{MN} \delta_{KL} + \bar{A}^{(1)}_{MN} C_{KL} - \\
&\quad - \hat{\gamma}_{-1}(C^{(-1)}_{KM} C^{(-1)}_{LN} + C^{(-1)}_{KN} C^{(-1)}_{LM}) + \\
&\quad + \hat{\gamma}_1(\delta_{KM}\delta_{LN} + \delta_{KN}\delta_{LM}), \\
B_{KL} &= -\left(\frac{\partial\hat{\gamma}_{-1}}{\partial T} C^{(-1)}_{KL} + \frac{\partial\hat{\gamma}_o}{\partial T}\delta_{KL} + \frac{\partial\hat{\gamma}_1}{\partial T} C_{KL}\right),
\end{aligned} \tag{4.3.5}$$

where

$$\bar{A}^{(i)}_{MN} = 2\left(\frac{\partial\hat{\gamma}_i}{\partial I_1}\delta_{MN} + \frac{\partial\hat{\gamma}_i}{\partial I_2}\Lambda_{MN} + \frac{\partial\hat{\gamma}_i}{\partial I_3} I_3 C^{(-1)}_{MN}\right) \quad (i = -1,0,1).$$

By (4.2.19) and (4.3.5),

$$\begin{aligned}
C_{ijrs} &= \delta_{ij}\hat{\alpha}^{(-1)}_{rs} + b_{ij}\hat{\alpha}^{(o)}_{rs} + b_{ip}b_{jp}\hat{\alpha}^{(1)}_{rs} - \\
&\quad - \hat{p}(\delta_{ir}\delta_{js} + \delta_{is}\delta_{jr}) - \hat{\Psi}(b_{ir}b_{js} + b_{is}b_{jr}), \\
\beta_{ij} &= -\frac{1}{J}\left(\frac{\partial\hat{\gamma}_{-1}}{\partial T}\delta_{ij} + \frac{\partial\hat{\gamma}_o}{\partial T} b_{ij} + \frac{\partial\hat{\gamma}_1}{\partial T} b_{is}b_{js}\right),
\end{aligned} \tag{4.3.6}$$

where

$$\hat{\alpha}_{rs}^{(i)} = \frac{2}{J}\left(\frac{\partial\hat{\gamma}_i}{\partial I_1} b_{rs} + \frac{\partial\hat{\gamma}_i}{\partial I_2} B_{rs} + I_3 \frac{\partial\hat{\gamma}_i}{\partial I_3} \delta_{rs}\right) \quad (i = -1,0,1).$$

If we define

$$\bar{A}_{ij} = \frac{2}{J} \frac{\partial^2\psi}{\partial I_i \partial I_j}, \quad L_i = \frac{2}{J} \frac{\partial^2\psi}{\partial I_i \, \partial T}, \tag{4.3.7}$$

then

$$\beta_{ij} = -I_3 L_3 \, \delta_{ij} - (L_1 + I_1 L_2) b_{ij} + L_2 b_{is} b_{js}, \tag{4.3.8}$$

and $\hat{\alpha}_{rs}^{(i)}$ assume expressions similar to that of $\alpha_{rs}^{(i)}$ in (1.3.20).

For an isotropic body we have (Pipkin and Rivlin [292], Green and Adkins [135], Sect. 8.6, Eringen and Şuhubi [116], Sect.1.14)

$$\hat{Q}_A = (K_1 \, \delta_{AB} + K_2 E_{AB} + K_3 E_{AM} E_{MB}) T_{,B}, \tag{4.3.9}$$

where K_i are functions of the invariants

$$I_1, I_2, I_3, I_4 = T_{,A} T_{,A}, I_5 = T_{,A} T_{,B} E_{AB}, I_6 = T_{,A} T_{,B} E_{BC} E_{CA}, T. \tag{4.3.10}$$

It follows from (4.2.26) and (4.3.9) that

$$\begin{aligned}
R_{KMN} &= R_{MN}^{(1)} T_{,K} + R_{MN}^{(2)} E_{KL} T_{,L} + R_{MN}^{(3)} E_{KA} E_{AL} T_{,L} + \\
&\quad + \frac{1}{2} K_2 (\delta_{KM}\delta_{LN} + \delta_{KN}\delta_{LM}) T_{,L} + \\
&\quad + \frac{1}{2} K_3 \, (\delta_{KM} E_{NL} + \delta_{LN} E_{KM} + \delta_{KN} E_{MN} + \delta_{ML} E_{KN}) T_{,L}, \\
D_K &= (D^{(1)} \, \delta_{KL} + D^{(2)} E_{KL} + D^{(3)} E_{KM} E_{ML}) T_{,L}, \\
K_{MN} &= (J_N^{(1)} \, \delta_{ML} + J_N^{(2)} E_{ML} + J_N^{(3)} E_{MA} E_{AL}) T_{,L} + \\
&\quad + K_1 \, \delta_{MN} + K_2 E_{MN} + K_3 E_{MA} E_{AN},
\end{aligned} \tag{4.3.11}$$

where

$$R_{MN}^{(i)} = 2\frac{\partial K_i}{\partial I_1}\delta_{MN} + 2\frac{\partial K_i}{\partial I_2}\Lambda_{MN} + 2I_3C_{MN}^{(-1)}\frac{\partial K_i}{\partial I_3} +$$

$$+ T_{,M}T_{,N}\frac{\partial K_i}{\partial I_5} + (E_{MA}T_{,N} + E_{NA}T_{,M})T_{,A}\frac{\partial K_i}{\partial I_6},$$

$$D^{(i)} = \frac{\partial K_i}{\partial T}, \quad J_N^{(i)} = 2(T_{,N}\frac{\partial K_i}{\partial I_4} + E_{NA}T_{,A}\frac{\partial K_i}{\partial I_5} + \tag{4.3.12}$$

$$+ E_{NA}E_{AL}T_{,L}\frac{\partial K_i}{\partial I_6}).$$

Using (4.2.33) and (4.3.11) we can find the expressions for h_{irs}, a_i and k_{ij}. When the primary state is at constant temperature we obtain

$$Q_A = 0, \quad R_{KMN} = 0, \quad D_K = 0,$$

$$K_{MN} = K_1\delta_{MN} + K_2E_{MN} + K_3E_{MA}E_{AN} = K_{NM}, \tag{4.3.13}$$

where K_i are evaluated for the constant temperature T.

4.4 Thermoelastic deformations with initial stress and initial heat flux

In this section we suppose that the primary state is identical with that of the initial body B_o so that $x_1 = X_1$, $x_2 = X_2$, $x_3 = X_3$. We assume that B_o is subjected to initial stress and is at nonuniform temperature $\bar{T}$. We then consider that the secondary state is obtained from B by an infinitesimal thermoelastic deformation. In this case,

$$J = 1, \quad E_{AB} = 0, \quad T = \bar{T}, \quad T_{AB} = \frac{\partial\hat{\psi}}{\partial E_{AB}},$$

$$\rho_o\eta = -\frac{\partial\hat{\psi}}{\partial T}, \quad Q_A = \hat{Q}_A(0,\bar{T},\bar{T}_{,A}), \tag{4.4.1}$$

the derivatives being evaluated at $E_{AB} = 0$ and $T = \bar{T}$. Clearly, T_{AB}, T_{Ai} and T_{ij} coincide. Moreover, $Q_A = q_i\,\delta_{iA}$. The initial stresses satisfy the conditions that the body B is in equilibrium. The initial heat flux Q_A is an arbitrary function of $\underline{X}$ apart from satisfying the energy equation.

It follows from (4.2.19), (4.2.24) and (4.2.33) that C_{ijrs}, β_{ij},

a, h_{irs}, a_i, k_{ij} are identical with A_{KLMN}, B_{KL}, A, R_{MNL}, D_K, K_{AB}, respectively. The coefficients defined by (4.2.11) and (4.2.26) are now evaluated at $E_{AB} = 0$ and $T = \bar{T}$.

The equations governing the infinitesimal thermoelastic deformations are the equations of motion

$$s_{ji,j} + \rho F_i = \rho \ddot{u}_i \text{ on } B \times (t_o, t_1), \tag{4.4.2}$$

the energy equation

$$\bar{T}\dot{\gamma} = \phi_{i,i} + \rho S \quad \text{on } B \times (t_o, t_1), \tag{4.4.3}$$

and the constitutive equations

$$\begin{aligned} s_{ij} &= (C_{ijrs} + t_{is}\,\delta_{jr})u_{r,s} - \beta_{ij}\theta, \\ \gamma &= \beta_{ij}\, u_{i,j} + a\theta, \\ \phi_i &= h_{irs}\, u_{r,s} + a_i\theta + k_{ij}\,\theta_{,j}. \end{aligned} \tag{4.4.4}$$

If $\bar{T}$ is constant, then $q_i = 0$, $h_{irs} = 0$, $a_i = 0$, and the theory reduces to that established by Green [137].

To the system of equations (4.4.2)-(4.4.4) we adjoin the boundary conditions (4.2.43) and the initial conditions (4.2.44). When the body B is isotropic, then (4.2.33), (4.3.6) and (4.3.12) yield

$$\begin{aligned} C_{ijrs} &= \lambda\delta_{ij}\delta_{rs} + \mu(\delta_{ir}\delta_{js} + \delta_{is}\delta_{jr}), \quad \beta_{ij} = \beta\delta_{ij}, \\ h_{irs} &= h_1\bar{T}_{,i}\delta_{rs} + h_2\bar{T}_{,r}\bar{T}_{,s}\bar{T}_{,i} + h_3(\delta_{ir}\delta_{ps} + \delta_{is}\delta_{rp})\bar{T}_{,p}, \\ a_i &= m\bar{T}_{,i}\,, \quad k_{ij} = k_o\bar{T}_{,i}\bar{T}_{,j} + k\delta_{ij}, \quad t_{ij} = p\delta_{ij}, \end{aligned} \tag{4.4.5}$$

where λ, μ, β, h_i, m, k_o, k and p are given functions.

The form of the free energy ψ^* may be written down immediately as a special case of (4.2.45) when $E_{AB} = 0$, $T = \bar{T}$, and is

$$\psi^* = \psi_o + t_{ij}e_{ij} - f\theta + \frac{1}{2}(C_{ijrs}e_{ij}e_{rs} - 2\beta_{ij}e_{ij} - - a\theta^2 + t_{ij}u_{r,i}u_{r,j}), \tag{4.4.6}$$

where the coefficients $\psi_o, t_{ij}, f, C_{ijrs}$, β_{ij} and a are, in general, functions of the coordinates x_i.

4.5 Infinitesimal initial deformations

In this section we consider the case in which the primary state is obtained from B_0 by a small thermoelastic deformation. In view of (4.2.11), the free energy ψ takes the form

$$\psi = \psi_o - \eta_o(T-T_0) + \frac{1}{2} c_{ABCD}E_{AB}E_{CD} - b_{AB}E_{AB}(T-T_0) - - \frac{1}{2} a_o(T-T_0)^2 + \frac{1}{6} c_{ABCDMN}E_{AB}E_{CD}E_{MN} - - \frac{1}{2} b_{ABCD}E_{AB}E_{CD}(T-T_0) - \frac{1}{2} d_{AB}E_{AB}(T-T_0)^2 - \frac{1}{6} h(T-T_0)^3. \tag{4.5.1}$$

The coefficients which occur in (4.5.1) are constants if the body B_0 is homogeneous. Clearly, these coefficients satisfy the symmetry relations

$$c_{ABCD} = c_{CDAB} = c_{BACD}, \; b_{AB} = b_{BA}, \; d_{AB} = d_{BA}, c_{ABCDMN} = c_{BACDMN} = c_{ABDCMN} = c_{ABCDNM} = c_{CDABMN} = = c_{ABMNCD} = c_{MNCDAB}, \; b_{ABCD} = b_{BACD} = b_{CDAB}. \tag{4.5.2}$$

In writing (4.5.1) we have used the assumption that B_0 is unstressed. We now let

$$v_i = x_i - \delta_{iA}X_A, \; \zeta = T - T_0, \tag{4.5.3}$$

and assume that $v_i = \varepsilon'' v_i'$, $\zeta = \varepsilon''\zeta'$ where ε'' is a constant small enough for squares and higher powers to be neglected and v_i' and ζ' are independent of ε''.

If we introduce (4.5.1) into (4.1.15) and (4.2.11), and retain only terms of order ε", we obtain

$$T_{AB} = c_{ABCD}\,\varepsilon_{CD} - b_{AB}\zeta,$$

$$A_{KLMN} = c_{KLMN} + c_{KLMNRS}\,\varepsilon_{RS} - b_{KLMN}\,\zeta, \tag{4.5.4}$$

$$B_{MN} = b_{MN} + b_{MNRS}\,\varepsilon_{RS} + d_{MN}\zeta,$$

$$A = a_o + d_{KL}\,\varepsilon_{KL} + h\zeta,$$

where

$$\varepsilon_{KL} = \frac{1}{2}(v_{K,L} + v_{L,K}). \tag{4.5.5}$$

Since the coefficients R_{AMK}, D_A and K_{MN} in (4.2.26) are found from first derivatives of Q_A corresponding to the primary state, we consider a quadratic expression for Q_A of the form

$$Q_A = \kappa_{AM}T_{,M} + g_{AKLM}E_{KL}T_{,M} + g_{AM}(T-T_o)T_{,M} + \frac{1}{2}\,\xi_{AMN}T_{,M}T_{,N}, \tag{4.5.6}$$

where the coefficients κ_{AM}, g_{AKLM}, g_{AM}, ξ_{AMN} are constants if the body is homogeneous. Clearly, the form (4.5.6) satisfies the condition (4.1.17). The coefficients g_{AKLM} and ξ_{AMN} have the symmetry properties

$$g_{AKLM} = g_{ALKM}, \quad \xi_{AMN} = \xi_{ANM}. \tag{4.5.7}$$

It follows from (4.2.26), (4.5.3) and (4.5.6) that

$$R_{AMN} = g_{AMNK}\,\zeta_{,K}, \quad D_A = g_{AM}\,\zeta_{,M},$$

$$K_{MN} = \kappa_{MN} + g_{MKLN}\,\varepsilon_{KL} + g_{MN}\zeta + \xi_{MNK}\,\zeta_{,K}. \tag{4.5.8}$$

By (4.2.16), (4.2.29), (4.5.3), (4.5.4) and (4.5.8),

$$E_{iAjN} = c_{ABMN}(\delta_{iB}\delta_{jM} + \delta_{jM}v_{i,B} + \delta_{iB}v_{j,M}) +$$

$$+ (c_{ABMNRS}\,\varepsilon_{RS} - b_{ABMN})\delta_{iB}\delta_{jM}, \tag{4.5.9}$$

$$F_{Ai} = b_{AK}(\delta_{iK} + v_{i,K}) + (b_{AKRS}\varepsilon_{RS} + d_{AK}\zeta)\delta_{iK},\; G_{AiK} = g_{AMKL}\zeta_{,L}\delta_{iM}.$$

We now suppose that the continuum in the state B_o is isotropic. Then, we have

$$\begin{aligned} b_{KLMN} &= \beta_1\delta_{KL}\delta_{MN} + \beta_2(\delta_{KM}\delta_{LN} + \delta_{KN}\delta_{LM}), \\ g_{KLMN} &= k_1\delta_{KL}\delta_{MN} + k_2(\delta_{KM}\delta_{LN} + \delta_{KN}\delta_{LM}), \\ b_{KL} &= \beta\delta_{KL},\; d_{KL} = d\delta_{KL},\; g_{KL} = g\delta_{KL}, \\ \kappa_{MN} &= k\delta_{MN},\; \xi_{MNR} = 0, \end{aligned} \tag{4.5.10}$$

and c_{KLMN} and c_{KLMNRS} are given by (1.5.8).

It follows from (4.5.4), (4.5.8)-(4.5.10) and (1.5.8) that

$$\begin{aligned} E_{iAjN} &= (\lambda + \nu_1\varepsilon_{RR} - \beta_1\zeta)\delta_{iA}\delta_{jN} + (\mu + \nu_2\varepsilon_{RR} - \\ &- \beta_2\zeta)(\delta_{jA}\delta_{iN} + \delta_{ij}\delta_{AN}) + (\lambda v_{M,N} + 2\nu_2\varepsilon_{MN})\delta_{iA}\delta_{jM} + \\ &+ (\lambda v_{L,A} + 2\nu_2\varepsilon_{AL})\delta_{iL}\delta_{jN} + (\mu v_{L,N} + 2\nu_3\varepsilon_{LN})\delta_{jA}\delta_{iL} + \\ &+ (\mu v_{M,A} + 2\nu_3\varepsilon_{AM})\delta_{iN}\delta_{jM} + 2(\mu + \nu_3)\delta_{AN}\delta_{iP}\delta_{jQ}\varepsilon_{PQ} + \\ &+ 2\nu_3\delta_{ij}\varepsilon_{AN}, \\ T_{AB} &= \lambda\varepsilon_{RR}\delta_{AB} + 2\mu\varepsilon_{AB} - \beta\zeta\delta_{AB}, \\ F_{Ai} &= (\beta + \beta_1\varepsilon_{RR} + d\zeta)\delta_{iA} + \beta\delta_{iM}v_{M,A} + 2\beta_2\delta_{iL}\varepsilon_{AL}, \\ A &= a_o + d\varepsilon_{RR} + h\zeta,\quad D_A = g\zeta_{,A}, \end{aligned} \tag{4.5.11}$$

$$G_{AiK} = k_1\delta_{iK}\zeta_{,A} + k_2(\delta_{iA},\zeta_{,K} + \delta_{AK}\delta_{iM}\zeta_{,M}),$$

$$K_{MN} = (k + k_2\varepsilon_{RR} + g\zeta)\delta_{MN} + (k_1 + k_2)\varepsilon_{MN}.$$

By (1.5.8), (4.2.19), (4.2.24), (4.2.33), (4.5.3), (4.5.4), (4.5.8) and (4.5.10) we find

$$\begin{aligned} C_{ijrs} &= [\lambda+(\nu_1-\nu)\varepsilon_{mn}-\beta_1\zeta]\delta_{ij}\delta_{rs} + [\mu+(\nu_2-\mu)\varepsilon_{mn}-\beta_2\zeta](\delta_{ir}\delta_{js} + \\ &\quad + \delta_{is}\delta_{jr}) + 2(\lambda+\nu_2)(\delta_{ij}\varepsilon_{rs} + \delta_{rs}\varepsilon_{ij}) + \\ &\quad + 2(\mu+\nu_3)(\delta_{ir}\varepsilon_{js} + \delta_{js}\varepsilon_{ir} + \delta_{is}\varepsilon_{jr} + \delta_{jr}\varepsilon_{is}), \\ \beta_{ij} &= [\beta+(\beta_1-\beta)\varepsilon_{rr}+d\zeta]\delta_{ij}+2(\beta+\beta_2)\varepsilon_{ij}, \qquad (4.5.12) \\ a &= a_o + (d-a_o)\varepsilon_{mn} + h\zeta, \quad a_i = g\delta_{iA}\zeta_{,A}, \\ h_{ijk} &= k_1\delta_{jk}\delta_{iA}\zeta_{,A} + k_2(\delta_{ij}\delta_{kM}\zeta_{,M} +\delta_{ik}\delta_{jN}\zeta_{,N}), \\ k_{ij} &= k(\delta_{ij} + \delta_{iB}v_{j,B} + \delta_{jA}v_{i,A} - \delta_{ij}\varepsilon_{RR}) + \\ &\quad + (k_2\varepsilon_{RR} + g\zeta)\delta_{ij} + (k_1 + k_2)\varepsilon_{ij}, \end{aligned}$$

where $\delta_{ij} = \delta_{iA}\delta_{jB}\varepsilon_{AB}$. The case when the first deformed state is an isothermal one was studied by Green [137]. The second-order thermoelastic coefficients have been studied by Chadwick and Set [54].

5 Thermoelastodynamics

5.1 Thermoelastic processes

The fundamental system of field equations describing the infinitesimal thermoelastic deformations of a strained body, referred to the configuration B_o, consists of the equations of motion

$$S_{Ai,A} + \rho_o F_i = \rho_o \ddot{u}_i, \tag{5.1.1}$$

the energy equation

$$T\dot{\gamma} + \rho_o \dot{\eta}\theta = \Phi_{A,A} + \rho_o S, \tag{5.1.2}$$

and the constitutive equations

$$S_{Ai} = D_{iAjK} u_{j,K} - F_{Ai}\theta, \quad \gamma = F_{Ai} u_{i,A} + A\theta,$$
$$\Phi_A = G_{AiK} u_{i,K} + D_A \theta + K_{AM}\, \theta_{,M}, \tag{5.1.3}$$

on $B_o \times (0,t_1)$. Here we have used the notation

$$D_{iAjN} = E_{iAjN} + \delta_{ij} T_{AN}, \tag{5.1.4}$$

and we assumed that $t_o = 0$. By (4.2.11) and (4.2.16),

$$D_{iAjB} = D_{jBiA}. \tag{5.1.5}$$

We assume that D_{iAjB}, F_{Ai}, A, G_{AiK}, D_A, K_{MN}, T, η, ρ_o, F_i and S are prescribed functions with the properties: (i) D_{iAjB}, F_{Ai}, A, G_{AiK}, D_A, K_{MN}, η and T are continuously differentiable on $\bar{B}_o \times [0,t_1)$; (ii) ρ_o is continuous and strictly positive on $\bar{B}_o$; (iii) D_{iAjB} satisfy the conditions (5.1.5); (iv) F_i and S are continuous on $\bar{B}_o \times [0,t_1)$.

We say that θ is an admissible temperature field on $\bar{B}_o \times [0,t_1)$ provided: (i) θ is of class $C^{2,1}$ on $B_o \times (0,t_1)$; (ii) θ, $\theta_{,A}$ and $\dot{\theta}$ are continuous on $\bar{B}_o \times [0,t_1)$. An admissible heat flux vector Φ_A on $\bar{B}_o \times [0,t_1)$ is defined by the following conditions: (i) Φ_A is of class $C^{1,0}$ on $B_o \times (0,t_1)$; (ii) Φ_A and $\Phi_{L,L}$ are continuous on $\bar{B}_o \times [0,t_1)$. By an admissible process on $\bar{B}_o \times [0,t_1)$ we mean an ordered array of functions $p = (u_i,\theta,S_{Aj},\Phi_K)$ with the following properties: (i) $\underline{u}$ is admissible displacement field on $\bar{B}_o \times [0,t_1)$, (cf. Sect. 3.1); (ii) θ is an admissible temperature field on $\bar{B}_o \times [0,t_1)$; (iii) S_{Ai} is an admissible stress field on $\bar{B}_o \times [0,t_1)$, (cf. Sect. 3.1); (iv) Φ_A is an admissible heat flux vector.

When we omit mention of the domain of definition of an admissible process, it will always be understood to be $\bar{B}_o \times [0,t_1)$. We say that $p=(u_i,\theta,S_{Aj},\Phi_K)$ is an thermoelastic process corresponding to the body force $\underline{F}$ and the heat supply S if p is an admissible process that satisfies the fundamental system of field equations (5.1.1)-(5.1.3) on $B_o \times (0,t_1)$. The corresponding surface traction $\underline{P}$ and the surface heat flux Φ are defined at every regular point of $\partial B_o \times [0,t_1)$ by

$$P_i = S_{Ai}N_A, \quad \Phi = \Phi_A N_A. \tag{5.1.6}$$

Given a thermoelastic process, we define the functions w and I on $[0,t_1)$ by

$$w = \frac{1}{2}\int_{B_0} (\rho_0 \dot{u}_i \dot{u}_i + D_{iAjK}u_{i,A}u_{j,K} + A\theta^2)dV,$$

$$I = w - \frac{1}{2}\int_0^t \int_{B_0} [\dot{D}_{iAjK}u_{i,A}u_{j,K} - \dot{A}\theta^2 - 2\dot{F}_{Ai}u_{i,A}\theta - \tag{5.1.7}$$

$$- \frac{2}{T}\rho_0 \dot{\eta}\theta^2 - 2(\frac{1}{T}\theta)_{,A}\Phi_A]dV\, d\tau.$$

<u>Theorem 5.1.1</u>. Let $(u_i,\theta, S_{Ai},\Phi_K)$ be a thermoelastic process corresponding to the body force $\underline{F}$ and the heat supply S. Then

$$\int_{\partial B_0} (P_i\dot{u}_i + \frac{1}{T}\Phi\theta)dA + \int_{B_0} \rho_0(F_i\dot{u}_i + \frac{1}{T}S\theta)dV = \dot{I}. \tag{5.1.8}$$

Proof. By (5.1.3) and (5.1.5),

$$S_{Ai}\dot{u}_{i,A} + \dot{\gamma}\theta = \frac{1}{2}\frac{\partial}{\partial t}(D_{iAjK}u_{i,A}u_{j,K} + A\theta^2) - \\ - \frac{1}{2}\dot{D}_{iAjK}u_{i,A}u_{j,K} + \frac{1}{2}\dot{A}\theta^2 + \dot{F}_{Ai}u_{i,A}\theta. \tag{5.1.9}$$

On the other hand, in view of (5.1.1) and (5.1.2) we can write

$$S_{Ai}\dot{u}_{i,A} + \dot{\gamma}\theta = (S_{Ai}\dot{u}_i + \frac{1}{T}\theta\Phi_A)_{,A} - S_{Ai,A}\dot{u}_i - (\frac{1}{T}\theta)_{,A}\Phi_A +$$

$$+ \frac{1}{T}\rho_0 S\theta - \frac{1}{T}\rho_0\dot{\eta}\theta^2 = (S_{Ai}u_i + \frac{1}{T}\theta\Phi_A)_{,A} + \rho_0(F_i\dot{u}_i + \frac{1}{T}S\theta) -$$

$$- \rho_0\dot{u}_i\ddot{u}_i - \frac{1}{T}\rho_0\dot{\eta}\theta^2 - (\frac{1}{T}\theta)_{,A}\Phi_A.$$

By the divergence theorem,

$$\int_{B_0}(S_{Ai}\dot{u}_{i,A} + \dot{\gamma}\theta)dV = \int_{\partial B_0}(P_i\dot{u}_i + \frac{1}{T}\Phi\theta)dA + \\ + \int_{B_0}\rho_0(F_i\dot{u}_i + \frac{1}{T}S\theta)dV - \int_{B_0}[\rho_0\dot{u}_i\ddot{u}_i + \\ + \frac{1}{T}\rho_0\dot{\eta}\theta^2 + (\frac{1}{T}\theta)_{,A}\Phi_A]dV. \tag{5.1.10}$$

The relations (5.1.7), (5.1.9) and (5.1.10) imply the desired result. □

The relation (5.1.8) can be written in the form

$$\int_0^t\int_{\partial B_0}(P_i\dot{u}_i + \frac{1}{T}\Phi\theta)dA\,d\tau + \int_0^t\int_{B_0}\rho_0(F_i\dot{u}_i + \frac{1}{T}S\theta)dV\,d\tau = \\ = I(t) - I(0). \tag{5.1.11}$$

The following proposition is an immediate consequence of Theorem 5.1.1.

Theorem 5.1.2. Let $(u_i, \theta, S_{Ai}, \Phi_K)$ be a thermoelastic process corresponding to the body force $\underline{F}$ and the heat supply S, and suppose that

$$\underline{F} = \underline{0},\ S = 0 \text{ on } \bar{B}_0 \times [0,t_1),\ P_i\dot{u}_i = 0,\ \Phi\theta = 0 \text{ on } \partial B_0 \times [0,t_1).$$

Then I is conserved, i.e.

$$I(t) = I(0), \quad 0 \leqq t < t_1. \tag{5.1.12}$$

We assume for the remainder of this section that the primary state is a configuration of equilibrium. We now consider that the deformation of the body is referred to the configuration B. In this case the fundamental system of field equations consists of the equations of motion

$$s_{ji,j} + \rho F_i = \rho \ddot{u}_i, \tag{5.1.13}$$

the energy equation

$$\frac{1}{J} T\dot{\gamma} = \phi_{i,i} + \rho S, \tag{5.1.14}$$

and the constitutive equations

$$\begin{gathered} s_{ji} = d_{ijrs}u_{r,s} - \beta_{ij}\theta,\ \gamma = J(\beta_{ij}u_{i,j} + a\theta), \\ \phi_i = h_{irs}u_{r,s} + a_i\theta + k_{ij}\,\theta_{,j}, \end{gathered} \tag{5.1.15}$$

on $B \times (0,t_1)$. Here we have used the notation

$$d_{ijrs} = C_{ijrs} + \delta_{ir}t_{js}. \tag{5.1.16}$$

We note that

$$d_{ijrs} = d_{rsij},\ \beta_{ij} = \beta_{ji}. \tag{5.1.17}$$

We assume that: (i) d_{ijrs}, β_{ij}, a, h_{irs}, a_i, k_{ij}, J and T are continuously differentiable on $\bar{B}$ and $d_{ijrs} = d_{rsij}$, $\beta_{ij} = \beta_{ji}$; (ii) ρ is continuous and strictly positive on $\bar{B}$; (iii) F_i and S are continuous on $\bar{B} \times [0,t_1)$. We now consider admissible processes defined on $\bar{B} \times [0,t_1)$. We say that

$p^0 = (u_i, \theta, s_{jk}, \phi_r)$ is a thermoelastic process on $\bar{B} \times [0,t_1)$ corresponding to the body force $\underline{F}$ and the heat supply S if p^0 is an admissible process on $\bar{B} \times [0,t_1)$ that satisfies (5.1.13)-(5.1.15) on $B \times (0,t_1)$. The corresponding surface traction $\underline{s}$ and the surface heat flux q are defined at every regular point of $\partial B \times [0,t_1)$ by

$$s_i = s_{ji}n_j, \quad q = \phi_i n_i. \tag{5.1.18}$$

It follows from (4.2.16), (4.2.19), (4.2.24), (4.2.33), (5.1.4) and (5.1.7) that

$$\begin{aligned} w &= \frac{1}{2} \int_B (\rho \dot{u}_i \dot{u}_i + d_{ijrs} u_{i,j} u_{r,s} + a\theta^2) dv, \\ I &= w + \int_0^t \int_B (\frac{1}{T}\,\theta)_{,i}\, \phi_i \; dv\; ds. \end{aligned} \tag{5.1.19}$$

If T is constant, then (4.2.34) implies that $\phi_i = k_{ij}\, \theta_{,j}$. Theorem 5.1.2 has the following immediate consequence.

<u>Theorem 5.1.3.</u> Let B be a configuration of equilibrium at constant temperature. Let $p^0 = (u_i, \theta, s_{jk}, \phi_r)$ be a thermoelastic process on $\bar{B} \times [0,t_1)$ and assume that

$$\underline{F} = 0,\ S = 0 \text{ on } \bar{B} \times [0,t_1),\ s_i \dot{u}_i = 0,\ q\theta = 0 \text{ on } \partial B \times [0,t_1). \tag{5.1.20}$$

Then

$$w(t) - w(0) = -\frac{1}{T} \int_0^t \int_B k_{ij}\, \theta_{,i}\, \theta_{,j} dv ds, \quad 0 \leq t < t_1. \tag{5.1.21}$$

The relation (5.1.21) was used in the classical thermoelasticity to establish uniqueness (see Weiner [368], V. Ionescu-Cazimir [204]).

5.2 The boundary-initial-value problems

We first assume that the primary state is a configuration of equilibrium and we take B as reference configuration. Then, the fundamental system of field equations is given by (5.1.13)-(5.1.15). In addition to the specifications of d_{ijrs}, β_{ij}, a, h_{irs}, a_i, k_{ij}, J, ρ and T, we assume that the following data are given: (i) the initial displacement, velocity and temperature $\tilde{\underline{a}}$, $\tilde{\underline{b}}$ and $\tilde{\theta}_0$, all continuous on $\bar{B}$; (ii) the surface displacement $\tilde{\underline{u}}$ continuous on $\bar{S}_1 \times [0,t_1)$; (iii) the surface traction $\tilde{\underline{p}}$ piecewise regular on $S_2 \times [0,t_1)$ and continuous in time; (iv) the surface temperature $\tilde{\theta}$ continuous on $\bar{S}_3 \times [0,t_1)$; (v) the surface heat flux $\tilde{\phi}$ piecewise regular on $\bar{S}_4 \times [0,t_1)$ and continuous in time; (vi) the body force $\underline{F}$ and heat supply S both continuous on $\bar{B} \times [0,t_1)$. By an *external data system* we mean an ordered array $L = (\underline{F}, S, \tilde{\underline{u}}, \tilde{\underline{p}}, \tilde{\theta}, \tilde{\phi}, \tilde{\underline{a}}, \tilde{\underline{b}}, \tilde{\theta}_0)$ with the properties (i)-(vi).

The mixed problem of the dynamic theory of thermoelasticity consists in finding a thermoelastic process $(u_i, \theta, s_{jk}, \phi_i)$ on $\bar{B} \times [0,t_1)$ corresponding to the body force $\underline{F}$ and heat supply S, that satisfies the initial conditions

$$\underline{u}(\cdot,0) = \tilde{\underline{a}}, \quad \dot{\underline{u}}(\cdot,0) = \tilde{\underline{b}}, \; \theta(\cdot,0) = \tilde{\theta}_0 \text{ on } \bar{B}, \tag{5.2.1}$$

and the boundary conditions

$$\begin{aligned} &\underline{u} = \tilde{\underline{u}} \text{ on } \bar{S}_1 \times [0,t_1), \quad \underline{s} = \tilde{\underline{p}} \text{ on } S_2 \times [0,t_1), \\ &\theta = \tilde{\theta} \text{ on } \bar{S}_3 \times [0,t_1), \quad q = \tilde{\phi} \text{ on } S_4 \times [0,t_1). \end{aligned} \tag{5.2.2}$$

Substitution of (5.1.15) into the equations (5.1.13), (5.1.14) and (5.2.2) yields the equations

$$\begin{aligned} &(d_{ijrs}u_{r,s})_{,j} - (\beta_{ij}\theta)_{,j} + \rho F_i = \rho\ddot{u}_i, \\ &(k_{ij}\theta_{,j} + a_i\theta)_{,i} + (h_{irs}u_{r,s})_{,i} - T\beta_{ij}\dot{u}_{i,j} - aT\dot{\theta} = \\ &\qquad = -\rho S, \end{aligned} \tag{5.2.3}$$

on $B \times (0,t_1)$ and the boundary conditions

$$\begin{aligned} &u_i = \tilde{u}_i \text{ on } \bar{S}_1 \times [0,t_1), \; (d_{ijrs}u_{r,s} - \beta_{ij}\theta)n_j = \tilde{p}_i \text{ on } S_2 \times [0,t_1), \\ &\theta = \tilde{\theta} \text{ on } \bar{S}_3 \times [0,t_1), \; (k_{ij}\theta_{,j} + a_i\theta + h_{irs}u_{r,s})n_i = \tilde{\phi} \text{ on } S_4 \times [0,t_1). \end{aligned} \tag{5.2.4}$$

We recall that if T is constant, then $h_{irs} = 0$ and $a_i = 0$.

By a solution of the mixed problem we mean the ordered pair $(\underline{u},\theta)$ with the following properties: (i) $\underline{u}$ is an admissible displacement field on $\bar{B} \times [0,t_1)$; (ii) θ is an admissible temperature field on $\bar{B} \times [0,t_1)$; (iii) $\underline{u}$ and θ satisfy the equations (5.2.3) on $B \times (0,t_1)$, the initial conditions (5.2.1) and the boundary conditions (5.2.4).

Let h_j and g be functions on $\bar{B} \times [0,t_1)$ defined by

$$h_j = \rho(i * F_j + \tilde{a}_j + t\tilde{b}_j),$$
$$g = 1 * \rho S + aT\tilde{\theta}_0 + T\beta_{ij}\tilde{a}_{ij}, \tag{5.2.5}$$

where the function i is given by (3.1.11).

Theorem 5.2.1. Let $p = (u_i,\theta,s_{jk},\phi_s)$ be an admissible process on $\bar{B} \times [0,t_1)$. Then p satisfies the equations of motion (5.1.13), the equation of energy (5.1.14) and the initial conditions (5.2.1) if and only if

$$i * s_{kj,k} + h_j = \rho u_j, \quad T\gamma = J(1 * \phi_{i,i} + g). \tag{5.2.6}$$

The proof of this theorem is analogous to that of Theorem 3.1.3 and can safely be omitted (see Carlson [47], Sect. 22 and Lebon [241], p. 370).

The next proposition is an immediate consequence of Theorem 5.2.1.

Theorem 5.2.2. Let $\underline{u}$ be an admissible displacement field on $\bar{B} \times [0,t_1)$, and let θ be an admissible temperature field on $\bar{B} \times [0,t_1)$. Then $\Lambda = (\underline{u},\theta)$ is a solution of the mixed problem if and only if it meets the field equations

$$i * (d_{kjrs}u_{r,s} - \beta_{kj}\theta)_{,k} + h_j = \rho u_j,$$
$$1 * (h_{irs}u_{r,s} + a_i\theta + k_{ij}\theta_{,j})_{,i} - T(\beta_{ij}u_{i,j} + a\theta) = -g, \tag{5.2.7}$$

on $B \times [0,t_1)$, and the boundary conditions (5.2.4).

We now assume that the primary state is a time-dependent configuration. Then we take B_0 as reference configuration. The fundamental system of field equations consists of the equations (5.1.1)-(5.1.3) on $B_0 \times (0,t_1)$. To this system we adjoin the initial conditions

$$\underline{u}(\cdot,0) = \hat{\underline{a}}, \quad \dot{\underline{u}}(\cdot,0) = \hat{\underline{b}}, \quad \theta(\cdot,0) = \hat{\theta}_o \text{ on } \bar{B}_o, \tag{5.2.8}$$

and the boundary conditions

$$\begin{aligned} &\underline{u} = \hat{\underline{u}} \text{ on } \bar{\Sigma}_1 \times [0,t_1), \quad \underline{P} = \hat{\underline{P}} \text{ on } \Sigma_2 \times [0,t_1), \\ &\theta = \hat{\theta} \text{ on } \bar{\Sigma}_3 \times [0,t_1), \quad \Phi = \hat{\Phi} \text{ on } \Sigma_4 \times [0,t_1). \end{aligned} \tag{5.2.9}$$

We assume that: (i) $\hat{\underline{a}}$, $\hat{\underline{b}}$ and $\hat{\theta}_o$ are continuous on $\bar{B}_o$; (ii) $\hat{\underline{u}}$ is continuous on $\bar{\Sigma}_1 \times [0,t_1)$; (iii) $\hat{\underline{P}}$ is continuous in time and piecewise regular on $\Sigma_2 \times [0,t_1)$; (iv) $\hat{\theta}$ is continuous on $\bar{\Sigma}_3 \times [0,t_1)$; (v) $\hat{\Phi}$ is continuous in time and piecewise regular on $\Sigma_4 \times [0,t_1)$; (vi) $\underline{F}$ and S are continuous on $\bar{B}_o \times [0,t_1)$. By an external data system on $\bar{B}_o \times [0,t_1)$ we mean an ordered array $L^o = (\underline{F}, S, \hat{\underline{u}}, \hat{\underline{P}}, \hat{\theta}, \hat{\Phi}, \hat{\underline{a}}, \hat{\underline{b}}, \hat{\theta}_o)$ with the properties (i)-(vi).

Substitution of (5.1.3) into the equations (5.1.1), (5.1.2) and (5.2.9) yields the equations

$$\begin{aligned} &(D_{iAjK}u_{j,K})_{,A} - (F_{Ai}\theta)_{,A} + \rho_o F_i = \rho_o \ddot{u}_i, \\ &(K_{AM}\,\theta_{,M})_{,A} + (D_A\theta + G_{AiK}u_{i,K})_{,A} - \rho_o\dot{\eta}\theta - T(F_{Ai}\dot{u}_{i,A} + \\ &\qquad + A\dot{\theta} + \dot{F}_{Ai}u_{i,A} + \dot{A}\theta) = -\rho_o S, \end{aligned} \tag{5.2.10}$$

and the boundary conditions

$$\begin{aligned} &u_i = \hat{u}_i \text{ on } \bar{\Sigma}_1 \times [0,t_1), \quad (D_{iAjK}u_{j,K} - F_{Ai}\theta)N_A = \hat{P}_i \text{ on} \\ &\qquad\qquad \Sigma_2 \times [0,t_1), \\ &\theta = \hat{\theta} \text{ on } \bar{\Sigma}_3 \times [0,t_1), \; (K_{AM}\theta_{,M} + G_{AiK}u_{i,K} + D_A\theta)N_A = \hat{\Phi} \\ &\qquad\qquad \text{on } \Sigma_4 \times [0,t_1). \end{aligned} \tag{5.2.11}$$

By a solution of the problem we mean the ordered pair $(\underline{u},\theta)$ with the properties: (i) $\underline{u}$ is an admissible displacement field on $\bar{B}_0 \times [0,t_1)$; (ii) θ is an admissible temperature field on $\bar{B}_0 \times [0,t_1)$; (iii) $\underline{u}$ and θ satisfy the equations (5.2.10) on $B_0 \times (0,t_1)$, the initial conditions (5.2.8) and the boundary conditions (5.2.11).

5.3 Uniqueness and continuous dependence results

α) Isothermal primary states. We assume that the primary state is a configuration of equilibrium at constant temperature. The fundamental system of field equations is given by (5.1.13)-(5.1.15) with $h_{irs} = 0$ and $a_i \equiv 0$. To this system we adjoin the initial conditions (5.2.1) and the boundary conditions (5.2.2).

We first present the uniqueness and continuous dependence results established by Knops and Payne [226]. The behaviour of the body is studied on a finite time interval $[0,t_1]$. The method they use is a modification of the method of logarithmic convexity. Let us now impose the following conditions:

(i) $\rho \geqq \rho_m > 0$, $a \geqq a_o > 0$, where ρ_m and a_o are constants;

(ii) there exists finite constants M_1 and M_2 such that

$$\beta_{ij}\beta_{ij} \leq M_1^2, \quad \beta_{ir,r}\,\beta_{ij,j} \leq M_2^2;$$

(iii) k_{ij} is positive definite in the sense that there exists a positive constant k_o such that

$$\int_B k_{ij}\xi_i\xi_j dv \geqq k_o \int_B \xi_i\xi_i dv,$$

for all vectors ξ_i;

(iv) S_2 is empty or S_4 is empty.

Let $(\underline{u}^{(1)}, \theta^{(1)})$ and $(\underline{u}^{(2)}, \theta^{(2)})$ be solutions corresponding to the external data systems $L^{(1)} = (\underline{F}, S, \tilde{\underline{u}}, \tilde{\underline{p}}, \tilde{\theta}, \tilde{\Phi}, \tilde{\underline{a}}^{(1)}, \tilde{\underline{b}}^{(1)}, \tilde{\theta}_o^{(1)})$ and $L^{(2)} = (\underline{F}, S, \tilde{\underline{u}}, \tilde{\underline{p}}, \tilde{\theta}, \tilde{\Phi}, \tilde{\underline{a}}^{(2)}, \tilde{\underline{b}}^{(2)}, \tilde{\theta}_o^{(2)})$, respectively. We define $\underline{u} = \underline{u}^{(1)} - \underline{u}^{(2)}$, and $\theta = \theta^{(1)} - \theta^{(2)}$. Then, $(\underline{u},\theta)$ is a solution of the mixed problem of the dynamic theory of thermoelasticity corresponding to the external data system $L_0 = (0,0,0,0,0,0,\underline{\tilde{a}},\underline{\tilde{b}},\theta_0)$ where $\underline{\tilde{a}} = \underline{\tilde{a}}^{(1)} - \underline{\tilde{a}}^{(2)}$, $\underline{\tilde{b}} = \underline{\tilde{b}}^{(1)} - \underline{\tilde{b}}^{(2)}$, $\theta_0 = \tilde{\theta}_0^{(1)} - \tilde{\theta}_0^{(2)}$. This problem is denoted by (P_0). In order to study the dependence on the initial data of solutions to (P_0), we first present some lemmas.

Let E, V and Q be the functions on $[0,t_1)$ defined by

$$E = \frac{1}{2}\int_B (\rho\dot{u}_i\dot{u}_i + d_{ijrs}u_{i,j}u_{r,s})dv, \quad V = \int_B a\theta^2\, dv,$$
$$Q = \int_B k_{ij}\,\theta_{,i}\,\theta_{,j}\, dv. \tag{5.3.1}$$

Lemma 5.3.1. Let $(\underline{u},\theta)$ be a solution of the problem (P_0). Then

$$E(t) - E(0) = -\int_0^t \int_B \dot{u}_i(\beta_{ij}\theta)_{,j}\, dv\, ds,$$
$$V(t) - V(0) = -\frac{2}{T}\int_0^t Q\, d\tau + 2\int_0^t \int_B \dot{u}_i(\beta_{ij}\theta)_{,j}\, dv\, ds. \tag{5.3.2}$$

Proof. In view of (5.1.13) and (5.1.14),

$$(s_{ji}\dot{u}_i)_{,j} - s_{ji}\dot{u}_{i,j} = \rho\dot{u}_i\ddot{u}_i, \quad (\phi_i\theta)_{,i} - \phi_i\,\theta_{,i} = \frac{1}{J}\, T\dot{\gamma}\theta. \tag{5.3.3}$$

By the divergence theorem, (5.3.3) and (5.1.15) we obtain

$$\int_{\partial B} (s_{ji} + \beta_{ij}\theta)\dot{u}_i n_j da = \dot{E} + \int_B (\beta_{ij}\theta)_{,j}\dot{u}_i dv,$$
$$\int_{\partial B} (\phi_j - T\beta_{ij}\dot{u}_i)n_j\theta da = \int_B [Ta\theta\dot{\theta} + \phi_i\,\theta_{,i} - T(\beta_{ij}\theta)_{,j}\dot{u}_i]dv. \tag{5.3.4}$$

The relations (5.3.4) and the property (iv) imply the desired result. □

We note that

$$\int_0^t (\int_0^s f(r)dr)ds = \int_0^t (t-s)f(s)ds = i * f, \tag{5.3.5}$$

where the function i is given by (3.1.11).

It follows from (5.3.2) and (5.3.5) that

$$1 * V = tV(0) - \frac{2}{T} i * Q + 2\int_0^t \int_B (t-s)\dot{u}_i(\beta_{ij}\theta)_{,j}\, dvds. \tag{5.3.6}$$

Let K be the function on $[0,t_1)$ defined by

$$K = \int_B \rho \dot{u}_i \dot{u}_i dv. \tag{5.3.7}$$

Lemma 5.3.2. Let $(\underline{u},\theta)$ be a solution of the problem (P_o). Then

$$1 * V + \frac{2}{T} i * Q \leqq 2tV(0) + 2(d_1 + d_2 t)(i * K), \tag{5.3.8}$$

$$V + \frac{1}{T} * Q \leqq (1+\nu t)V(0)+[(1+\nu t)d_1+(\nu^{-1}+\nu t^2)d_2](1 * K), \tag{5.3.9}$$

where ν is an arbitrary positive constant, and

$$d_1 = \frac{M_1^2 T}{k_o \rho_m}, \quad d_2 = \frac{2M_2^2}{a_o \rho_m}. \tag{5.3.10}$$

Proof. For future use we record that for real numbers a and b we have the inequality

$$ab \leqq \frac{1}{2}(\alpha a^2 + \alpha^{-1} b^2), \tag{5.3.11}$$

where α is an arbitrary positive constant. Applying this inequality to the last term of (5.3.6) yields

$$I_1 \equiv 2\int_0^t \int_B (t-s)\dot{u}_i(\beta_{ij}\theta)_{,j} dvds \leqq \alpha_1 a_o^{-1}(1 * V) + \alpha_2 k_o^{-1}(i * Q) +$$

$$+ (\alpha_1^{-1} t M_2^2 + \alpha_2^{-1} M_1^2)\rho_m^{-1}(i * K),$$

for arbitrary positive constants α_1 and α_2. Here, we have made use of the conditions (i)-(iii), and the estimate

$$\int_0^t (t-s)^2 f^2(s)ds \leqq t \int_0^t (t-s)f^2(s)ds.$$

If we choose $2\alpha_1 = a_o$, $\alpha_2 = k_oT^{-1}$, then we obtain

$$I_1 \leq \frac{1}{2} * V + \frac{1}{T} i * Q + (d_1 + d_2t)(i * K). \qquad (5.3.12)$$

The relations (5.3.6) and (5.3.12) imply the formula (5.3.8). The following inequality is an immediate consequence of (5.3.8)

$$1 * V + \frac{2}{T} i * Q \leq 2tV(0) + 2(d_1t + d_2t^2)(1 * K). \qquad (5.3.13)$$

Using the inequality (5.3.11), we find that

$$2\int_0^t \int_B \dot{u}_i(\beta_{ij}\theta)_{,j}dvds \leq \beta_1 a_o^{-1}(1 * V) + (\beta_2 k_o)^{-1}(1 * Q) + \\ + (\beta_1^{-1}M_2^2 + \beta_2 M_1^2)\ \rho_m^{-1}(1 * K), \qquad (5.3.14)$$

where β_α are arbitrary positive constants. We choose $2\beta_1 = a_o\nu$, $\beta_2 k_o = T$, where ν is an arbitrary positive constant. Then (5.3.2), (5.3.13) and (5.3.14) imply (5.3.9). □

Let G be the function on $[0,t_1)$ defined by

$$G = \int_B \rho u_i\dot{u}_i dv. \qquad (5.3.15)$$

<u>Lemma 5.3.3</u>. If $(\underline{u},\theta)$ is a solution of the problem (P_o), then

$$G(t) - G(0) = 2 * (K - E) - \int_0^t \int_B u_i(\beta_{ij}\theta)_{,j}\ dvds. \qquad (5.3.16)$$

<u>Proof</u>. Since $\overline{u_i\dot{u}_i}^{\bm\cdot} = \dot{u}_i\dot{u}_i + u_i\ddot{u}_i$, it follows that

$$G(t) - G(0) = 1 * K + \int_0^t \int_B \rho u_i\ddot{u}_i\ dvds. \qquad (5.3.17)$$

On the other hand, in view of the equations of motion,

$$\rho u_i\ddot{u}_i = u_i s_{ji,j} = (u_i s_{ji})_{,j} - s_{ji}u_{i,j}. \qquad (5.3.18)$$

By (5.3.2), (5.3.18), the divergence theorem and the condition (iv), we arrive at

$$\int_B \rho u_i \ddot{u}_i dv = - \int_B [d_{ijrs} u_{i,j} u_{r,s} + u_i(\beta_{ij}\theta)_{,j}]dv. \tag{5.3.19}$$

The desired result follows from (5.3.1), (5.3.17) and (5.3.19). □

We define the function H on $[0,t_1)$ by

$$H = \int_B \rho u_i u_i \, dv. \tag{5.3.20}$$

Lemma 5.3.4. Let F be the function on $[0,t_1)$ defined by

$$F = 1 * H + (t_1 - t)H(0) + \Gamma, \tag{5.3.21}$$

where Γ is a non-negative constant which will be determined later. Then,

$$\dot{F} = H - H(0) = 2 * G, \tag{5.3.22}$$

$$\ddot{F} = 2G = 4 * K - 2L, \tag{5.3.23}$$

where

$$L = 2tE(0) + tV(0) - G(0) - 1 * V - \frac{2}{T} i * Q + \\ + \int_0^t \int_B u_i(\beta_{ij}\theta)_{,j} \, dv \, ds. \tag{5.3.24}$$

Proof. Clearly, (5.3.22) follows from (5.3.15), (5.3.20) and (5.3.21). The relations (5.1.19), (5.1.21) imply that

$$2 * E = 2tE(0) + tV(0) - 1 * V - \frac{2}{T} i * Q. \tag{5.3.25}$$

The remainder of proof follows at once from (5.3.16), (5.3.22) and (5.3.25).□

Using (5.3.11) and Schwarz's inequality, we can write

$$\int_0^t \int_B u_i(\beta_{ij}\theta)_{,j} dv d\eta \leq 2 * V + \frac{1}{16} d_2 * H +$$

$$+ (T^{-1}d_1)^{1/2}[(1 * H)(1 * Q)]^{1/2}. \qquad (5.3.26)$$

By (iii), (5.3.24) and (5.3.26),

$$L \leqq 2tE(0) + tV(0) - G(0) + 1 * V - \frac{2}{T} i * Q + \frac{1}{16} d_2 * H + \\ + (T^{-1}d_1)^{1/2}[(1 * H)(1 * Q)]^{1/2}. \qquad (5.3.27)$$

It follows from (5.3.8), (5.3.9) and (5.3.27) that

$$L \leqq 2tE(0) + 3tV(0) - G(0) + 2(d_1 + d_2t)(i * K) + \\ + \frac{1}{16} d_2 * H + d_1^{1/2}(1 * H)^{1/2}\{[(1 + \nu t)V(0)]^{1/2} + \\ + [(1 + \nu t)d_1 + (\nu^{-1} + \nu t^2)d_2]^{1/2}(1 * K)^{1/2}\}.$$

Clearly,

$$[d_1^{1/2}(1 * H)^{1/2}][(1 + \nu t)V(0)]^{1/2} \leqq \frac{1}{2}[d_1(1 * H) + (1 + \nu t)V(0)],$$

so that

$$L \leqq L_1 + 2(d_1+d_2t)(i * K) + \frac{1}{16} (8d_1+d_2)(1 * H) + \\ + \{d_1[(1+\nu t)d_1 + (\nu^{-1} + \nu t^2)d_2]\}^{1/2}[(1 * H)(1 * K)]^{1/2}. \qquad (5.3.28)$$

Here we have used the notation

$$L_1 = 2tE(0) + [3t + \frac{1}{2} (1 + \nu t)]V(0) - G(0). \qquad (5.3.29)$$

Lemma 5.3.5. If $(\underline{u},\theta)$ is a solution of the problem (P_0) then

$$i * K \leqq \frac{1}{2} \dot{F} + \int_0^t (L_1-p\Gamma)\exp[q(t-s)]ds + \frac{1}{2}q(F-\Gamma)\exp(qt) + \\ + \frac{p}{q}F(\exp(qt)-1) + \frac{1}{q}(\frac{p}{q} + \frac{q}{2})[(qt-1)\exp(qt)+1]H(0), \qquad (5.3.30)$$

where

$$p = \frac{1}{16}\{(8d_1 + d_2) + 4d_1[(1 + \nu t_1)d_1 + (\nu^{-1} + \nu t_1^2)d_2]\},$$
$$q = 2(d_1 + d_2 t_1). \tag{5.3.31}$$

Proof. It follows from (5.3.11), (5.3.23) and (5.3.28) that

$$1 * K = \frac{1}{4}\ddot{F} + \frac{1}{2}L \leq \frac{1}{4}\ddot{F} + \frac{1}{2}L_1 + \frac{1}{2}g(i * K) + \frac{1}{2} * K + \frac{1}{2}f(1 * H), \tag{5.3.32}$$

where

$$f = \frac{1}{16}\{(d_2 + 8d_1) + 4d_1[(1 + \nu t)d_1 + (\nu^{-1} + \nu t^2)d_2]\},$$
$$g = 2(d_1 + d_2 t).$$

Clearly, $f \leq p$, $g \leq q$. From (5.3.32) we obtain

$$1 * K \leq \frac{1}{2}\ddot{F} + L_1 + q(i * K) + p(1 * H). \tag{5.3.33}$$

We note that

$$1 * K = \frac{d}{dt}(i * K).$$

Thus, the inequality (5.3.33) can be written in the form

$$\frac{d}{dt}[(i * K)\exp(-qt)] \leq [\frac{1}{2}\ddot{F} + p(1 * H) + L_1]\exp(-qt).$$

The above relation and (5.3.21) imply

$$i * K \leq \{\frac{1}{2}\int_0^t \ddot{F}(s)\exp(-qs)ds + p\int_0^t F(s)\exp(-qs)ds + \int_0^t (L_1 - p\Gamma)\exp(-qs)ds\}\exp(qt).$$

If we integrate by parts, we arrive at

$$i * K \leqq \frac{1}{2}\dot{F} - \frac{p}{q}F + \frac{p}{q}F(0)\exp(qt) +$$
$$+ \left(\frac{p}{q} + \frac{q}{2}\right)\int_0^t \dot{F}(s)\exp[q(t-s)]ds + \qquad (5.3.34)$$
$$+ \int_0^t (L_1 - p\Gamma)\exp[q(t-s)]ds.$$

Since $\dot{F} + H(0) = H \geqq 0$, we have

$$\int_0^t \dot{F}(s)\exp[q(t-s)]ds \leqq [F(t)-F(0)]\exp(qt) + \frac{1}{q}[1 + \qquad (5.3.35)$$
$$+ (qt-1)\exp(qt)]H(0).$$

Substitution of (5.3.35) in (5.3.34) and $F(0) \geqq \Gamma$ imply the desired result. □

We say that the displacement field u is of class N if it satisfies the boundedness condition

$$\int_0^{t_1}\int_B \rho u_i u_i \, dvds \leqq N^2,$$

where N is some prescribed constant.

Theorem 5.3.1. If $(\underline{u},\theta)$ is a solution of the problem (P_0) and $\underline{u} \in N$, then for finite time t_1 it is possible to determine explicit positive constants c_i $(i = 1,2,\ldots,6)$, N_1 such that

$$\int_0^t\int_B \rho u_i u_i dvdt \leqq c_1 N_1^{2\delta}\{c_2 H(0) + c_3 K(0) + c_4|E(0)| + c_5 V(0)\}^{1-\delta},$$
$$t \in [0,t_1),$$

where δ is given by

$$\delta[1 - \exp(-c_6 t_1)] = 1 - \exp(-c_6 t).$$

Proof. By (5.3.21), (5.3.22), (5.3.23), (5.3.28), (5.3.30) and (5.3.31),

$$F\ddot{F} - \dot{F}^2 \geqq 4r^2 - 2FL \geqq 4r^2 - 2F\{L_1 + q(i * K) + \frac{1}{16}(d_2+8d_1)(1 * H) + \frac{1}{2}[16p - (d_2+8d_1)]^{1/2}[(1 * H)(1 * K)]^{1/2}\},$$

where

$$r = \{(1 * H)(1 * K) - (1 * G)^2\}^{1/2}.$$

Using Lemma 5.3.5, we have

$$F\ddot{F} - \dot{F}^2 \geqq 4r^2 - 2FP - qF\dot{F} - \frac{1}{8}(8d_1+d_2)F^2 - q[q + 2\frac{p}{q}(1 - \exp(-qt_1))]F^2\exp(qt_1) - [16p - (8d_1+d_2)]^{1/2}F(r^2 + \dot{F}^2)^{1/2}, \tag{5.3.36}$$

where

$$P = L_1 + (\frac{p}{q} + \frac{q}{2})[(qt_1-1)\exp(qt_1) + 1]H(0) + q\int_0^t (L_1-p\Gamma)\exp[q(t-s)]ds - \frac{1}{2}q^2\Gamma\exp(qt_1) - \frac{1}{16}(8d_1+d_2)\Gamma.$$

Since,

$$|\dot{F}| \leqq H + H(0),$$

it follows that

$$|\dot{F}| \leqq \dot{F} + 2H(0).$$

Thus, we have

$$(r^2 + \dot{F}^2)^{1/2} \leqq r + |\dot{F}| \leqq r + \dot{F} + 2H(0). \tag{5.3.37}$$

Let us now take Γ of the form

$$\Gamma = b_1V(0) + b_2H(0) + b_3K(0) + b_4|E(0)|, \qquad (5.3.38)$$

where b_i $(i = 1,2,3,4)$ are chosen so large that

$$P - [16p - (8d_1+d_2)]^{1/2}H(0) \leqq 0. \qquad (5.3.39)$$

It follows from (5.3.36), (5.3.37) and (5.3.39) that

$$F\ddot{F} - \dot{F}^2 \geqq 4r^2 - m_1F^2 - m_2Fr - m_3F\dot{F}, \qquad (5.3.40)$$

where

$$m_1 = \frac{1}{8}(8d_1+d_2) + q[q + \frac{2p}{q}(1 - \exp(-qt_1))]\exp(qt_1).$$

$$m_2 = [16p - (8d_1+d_2)]^{1/2}, \quad m_3 = m_2 + q.$$

The inequality (5.3.40) implies

$$F\ddot{F} - \dot{F}^2 \geqq - 2p_1F^2 - p_2F\dot{F}, \qquad (5.3.41)$$

where

$$32p_1 = 16m_1 + m_2^2, \quad p_2 = m_3.$$

If $F = 0$ on $[0,t_1]$, then the problem is of no interest. We assume that there exists an interval (s_1,s_2) such that $F(t) > 0$, $t \in (s_1,s_1)$. If $t \in (s_1,s_2)$ then (5.3.41) implies

$$\frac{d^2}{dt^2}(\ln F) + p_2\frac{d}{dt}(\ln F) + 2p_1 \geqq 0.$$

Setting $y = \exp(-p_2t)$, the above relation becomes

$$\frac{d^2}{dy^2}\{\ln(F^{o}y^{b})\} \geqq 0,$$

where $F^{o}(y) = F[t(y)]$, and $b = -2p_1/p_2^2$. Thus, we conclude that $\ln(F^{o}y^{b})$ is a convex function. We obtain

$$F^{o}(y)y^{b} \leqq [F(y_1)y_1^{b}]^{v(y-y_2)}[F(y_2)y_2^{b}]^{v(y_1-y)}. \qquad (5.3.42)$$

where

$$y_\alpha = \exp(-p_2 s_\alpha) \quad (\alpha = 1,2), \quad v^{-1} = y_1 - y_2.$$

Now either $s_1 = 0$ or by continuity, $F(s_1) = 0$. When $F(s_1) = 0$, the relation (5.3.42) gives at once $F(t) = 0$ for $s_1 \leqq t \leqq s_2$ and hence for $0 \leqq t \leqq t_1$. This fact implies the uniqueness of u_i. We assume now that $s_1 = 0$. Then,

$$F(t)\exp(ct) \leqq [F(0)]^{g_1}[F(t_1)\exp(ct_1)]^{g_2}, \qquad (5.3.43)$$

where

$$c = 2p_1/p_2, \; g_1 = [\exp(-p_2 t) - \exp(-p_2 t_1)]/[1 - \exp(-p_2 t_1)],$$

$$g_2 = [1 - \exp(-p_2 t)]/[1 - \exp(-p_2 t_1)].$$

If $\underline{u} \in N$ and if the initial temperature, displacement and velocity are square integrable and the initial energy is bounded, then there exists a constant N_1 such that

$$F(t_1) \leqq N_1^2. \qquad (5.3.44)$$

It follows from (5.3.43) and (5.3.44) that

$$F(t) \leqq N_1^{2\delta}[t_1 H(0) + \Gamma]^{1-\delta} \exp[c(\delta t_1 - t)], \qquad (5.3.45)$$

where $\delta = g_2$. By (5.3.11), (5.3.38) and (5.3.45) we arrive at the desired result. □

The theorem does not establish continuous dependence on initial data of θ in any appropriate norm. To obtain this, we note that (5.3.9) implies

$$\int_B a\theta^2 dv \leq (1 + \nu t)V(0) + [(1 + \nu t)d_1 + + (\nu^{-1} + \nu t^2)d_2] \int_0^t \int_B \rho \dot{u}_i \dot{u}_i \, dvds. \tag{5.3.46}$$

The continuous dependence now follows for θ, from (5.3.46) and the inequality established in Theorem 5.3.1 but with u_i replaced by $\dot{u}_i$ in the latter. Clearly, now the initial data must be such that the displacement, its first and second derivatives are all initially square integrable.

An immediate consequence of the above results is the

Theorem 5.3.2. Assume that the hypotheses (i)-(iv) hold. Then the mixed problem has at most one solution.

Levine [245] has extended the results of Knops and Payne to the case in which d_{ijrs}, β_{ij}, k_{ij}, ρ and a depend on t. Results on stability have also been obtained by Dafermos [91], Slemrod and Infante [329], Brun [41] and Wilkes [378].

We next present continuous dependence results established by Wilkes [378]. Let W be the function on $[0,t_1)$ defined by

$$W = H + \frac{1}{T} \int_0^t \int_B k_{ij} w_i w_j \, dv \, ds, \tag{5.3.47}$$

where

$$w_i = 1 * \theta_{,i} + h_{,i}, \tag{5.3.48}$$

and h is a function independent of t which will be determined later. We assume that k_{ij} is symmetric. Clearly,

$$\dot{W} = 2G + \frac{1}{T} \int_B k_{ij} w_i w_j \, dv, \tag{5.3.49}$$

$$\ddot{W} = 2K + 2 \int_B \rho u_i \ddot{u}_i dv + \frac{2}{T} \int_B k_{ij} \theta_{,i} w_j dv. \tag{5.3.50}$$

It follows from (5.3.18), (5.3.19) and (5.3.50) that

$$\ddot{W} = 2K - 2\int_B d_{ijrs}u_{i,j}u_{r,s}dv + 2\int_B \beta_{ij}u_{i,j}\theta\, dv -$$

$$- \frac{2}{T}\int_B \theta(k_{ij}w_j)_{,i}\, dv. \tag{5.3.51}$$

In view of (5.1.12), (5.1.19) we can write

$$\ddot{W} = 4K + 2V + 2\int_B \beta_{ij}\theta u_{i,j}dv + \frac{4}{T} * Q - \frac{2}{T}\int_B \theta(k_{ij}w_j)_{,i}dv -$$

$$- 4I(0). \tag{5.3.52}$$

By (5.2.3),

$$(k_{ij}(1 * \theta_{,j}))_{,i} - T(\beta_{ij}u_{i,j} + a\theta) = -T(\beta_{ij}\tilde{a}_{i,j} + a\theta_o). \tag{5.3.53}$$

We define h to be a solution of the following boundary-value problem

$$(k_{ij}h_{,j})_{,i} = T(\beta_{ij}\tilde{a}_{i,j} + a\theta_o) \text{ on } B,$$

$$h = 0 \text{ on } \bar{S}_3, \quad k_{ij}h_{,j}n_i = 0 \text{ on } S_4. \tag{5.3.54}$$

If S_3 is not empty, then the solution of the problem (5.3.54) is unique. If S_3 is empty, then the problem (5.3.54) has a solution if and only if

$$\int_B (\beta_{ij}\tilde{a}_{i,j} + a\theta_o)dv = 0.$$

From (5.3.52)-(5.3.54) we obtain

$$\ddot{W} = 4K + \frac{4}{T} * Q - 4I(0). \tag{5.3.55}$$

We note that

$$k_{ij}w_iw_j - k_{ij}h_{,i}h_{,j} = 2\int_0^t k_{ij}\,\theta_{,i}w_j\, ds.$$

It then follows by Schwarz's inequality that

$$W\ddot{W} - (\dot{W} - \omega)^2 \geq -4I(0)W, \tag{5.3.56}$$

where

$$\omega = \frac{1}{T}\int_B k_{ij}h_{,i}h_{,j}dv.$$

First, we discuss continuous dependence for solutions with $I(0) \leqq 0$. Let R be the function on $[0,t_1]$ defined by

$$R = \ln[W + (t_1 - t)\omega]. \tag{5.3.57}$$

Next, by (5.3.55) and (5.3.56), $\ddot{R} \geqq 0$, so that

$$W(t) + (t_1-t)\omega \leqq [W(0) + t_1\omega]^{1-z}[W(t_1)]^z, \tag{5.3.58}$$

where $z = t/t_1$. We denote by Z the class of solutions to the problem (P_o) for which $W(t_1)$ is bounded. It follows from (5.3.58) that the solutions of class Z are continuously dependent upon the initial data on compact subintervals of $[0,t_1)$.

We now suppose that $I(0) > 0$. If we define R_1 by

$$R_1 = \ln[W + (t_1-t)\omega + 2I(0)] + t^2,$$

then (5.3.55) and (5.3.56) imply that

$$\ddot{R}_1 > 0.$$

We arrive at

$$W + (t_1-t)\omega + 2I(0) \leqq [W(0) + t_1\omega + 2I(0)]^{1-z}[W(t_1) + 2I(0)]^z \exp(tt_1-t^2),$$

which implies the continuous dependence on the initial data of the solutions belonging to Z.

β) <u>Time-dependent non-isothermal primary states.</u> We now present the uniqueness and continuous dependence results obtained by Chiriţa [77] for the case when the primary state is obtained from B_o by a non-isothermal motion. The results are obtained by a method based on a Gronwall-type inequality, which has been used by Dafermos [92] in the study of thermoelastic stability.

We consider the field equations (5.1.1)-(5.1.3), the initial conditions (5.2.8) and the boundary conditions (5.2.9). In what follows we examine the continuous dependence of the solutions upon the initial data and supply terms. Let $(\underline{u}^{(1)}, \theta^{(1)})$ and $(\underline{u}^{(2)}, \theta^{(2)})$ be solutions corresponding to the external data systems $D^{(1)} = (\underline{F}^{(1)}, S^{(1)}, \hat{\underline{u}}, \hat{\underline{P}}, \hat{\theta}, \hat{\Phi}, \hat{\underline{a}}^{(1)}, \hat{\underline{b}}^{(1)}, \hat{\theta}_o^{(1)})$and $D^{(2)} = (\underline{F}^{(2)}, S^{(2)}, \hat{\underline{u}}, \hat{\underline{P}}, \hat{\theta}, \hat{\Phi}, \hat{\underline{a}}^{(2)}, \hat{\underline{b}}^{(2)}, \hat{\theta}_o^{(2)})$, respectively. If we define $\underline{u} = \underline{u}^{(1)} - \underline{u}^{(2)}$ and $\theta = \theta^{(1)} - \theta^{(2)}$, then $(\underline{u},\theta)$ is a solution of the mixed problem corresponding to the external data system $L = (\underline{F},S,0,0,0,0,\hat{\underline{a}},\hat{\underline{b}},\theta_o)$ where $\underline{F} = \underline{F}^{(1)} - \underline{F}^{(2)}$, $S = S^{(1)} - S^{(2)}$, $\hat{\underline{a}} = \hat{\underline{a}}^{(1)} - \hat{\underline{a}}^{(2)}$, $\hat{\underline{b}} = \hat{\underline{b}}^{(1)} - \hat{\underline{b}}^{(2)}$, $\theta_o = \hat{\theta}_o^{(1)} - \hat{\theta}_o^{(2)}$. This problem is denoted by (P). We define the "distance" between the solutions $(\underline{u}^{(1)}, \theta^{(1)})$ and $(\underline{u}^{(2)}, \theta^{(2)})$ by

$$y = \frac{1}{2}\int_{B_o} (\rho_o \dot{u}_i \dot{u}_i + D_{iAjK}u_{i,A}u_{j,K} + A\theta^2)dV. \tag{5.3.59}$$

As before, the behaviour of the body is studied on the finite time interval $[0,t_1]$. We assume that

(i) $\rho_o \geqq \rho_1 > 0$, $A \geqq A_o > 0$, where ρ_1 and A_o are constants;

(ii) D_{iAjB} is positive definite in the sense that there exists a positive constant D_o such that

$$\int_{B_o} D_{iAjB}v_{iA}v_{jB}dV \geqq D_o \int_{B_o} v_{iA}v_{iA}dV,$$

for all tensors v_{iA} and any $t \in [0,t_1]$;

(iii) K_{MN} is positive definite in the sense that there exists a positive constant K_o such that

$$\int_{B_o} \frac{1}{T} K_{MN}\xi_M\xi_N \, dV \geq K_o \int_{B_o} \xi_M\xi_M \, dV,$$

for all vectors ξ_M and any $t \in [0,t_1]$;

(iv) Σ_2 and Σ_4 are empty.

The condition (iii) is the defining property of a definite elastic conductor introduced by Coleman and Gurtin [84]. The following proposition is an immediate consequence of Theorem 5.1.1.

Lemma 5.3.6. Let $(\underline{u},\theta)$ be a solution of the problem (P). Then

$$\dot{y} = \int_{B_o} \rho_o(F_i\dot{u}_i + \frac{1}{T} S\theta)dV + \frac{1}{2}\int_{B_o} (\dot{D}_{iAjK}u_{i,A}u_{j,K} - \dot{A}\theta^2 - 2\dot{F}_{Ki}u_{i,K}\theta - \frac{2}{T}\rho_o \dot{\eta}\theta^2)dV - \int_{B_o} \Phi_A(\frac{1}{T}\theta)_{,A} \, dV. \tag{5.3.60}$$

This lemma gives the evolution in time of the "distance" defined by (5.3.59).

Lemma 5.3.7. Let $(\underline{u},\theta)$ be a solution of the problem (P). Then there exists a positive constant c_o such that

$$-\int_{B_o} (\frac{1}{T}\theta)_{,A} \Phi_A dV \leq c_o \int_{B_o} (u_{i,K}u_{i,K} + \theta^2)dV. \tag{5.3.61}$$

Proof. By (5.1.3),

$$-\int_{B_o} (\frac{1}{T}\theta)_{,A} \Phi_A dV = -\int_{B_o} \frac{1}{T^2} \Phi_A(\theta_{,A}T - \theta T_{,A})dV = -\int_{B_o} \frac{1}{T}K_{AM}\theta_{,A}\theta_{,M}dV + \int_{B_o} (C\theta^2 + L_A\theta\theta_{,A} + V_{iK}u_{i,K}\theta + P_{AiK}u_{i,K}\theta_{,A})dV,$$

where

$$C = T^{-2}D_K T_{,K}, \quad L_M = T^{-2}(K_{AM}T_{,A} - TD_M),$$

$$V_{iK} = T^{-2}G_{AiK}T_{,A}, \quad P_{AiK} = -T^{-1}G_{AiK}.$$

Using (5.3.11) and Schwarz's inequality we obtain

$$- \int_{B_0} (\tfrac{1}{T}\theta)_{,A}\Phi_A dV \leqq - \int_{B_0} \tfrac{1}{T} K_{AM}\theta_{,A}\theta_{,M}dV + \tfrac{1}{2}(\alpha_1+\alpha_2)\int_{B_0} \theta_{,A}\theta_{,A}dV + \\ + \tfrac{1}{2}(m_1^2\alpha_1^{-1} + 2m_4^2 + 1)\int_{B_0} \theta^2 dV + \tfrac{1}{2}(m_2^2\alpha_2^{-1} + m_3^2)\int_{B_0} u_{i,A}u_{i,A}dV, \tag{5.3.62}$$

where α_ρ are arbitrary positive constants and

$$m_1^2 = \max(L_A L_A), \quad m_2^2 = \max(P_{AiK}P_{AiK}), \quad m_3^2 = \max(V_{iK}V_{iK}),$$

$$m_4^2 = \max|C| \text{ on } \bar{B}_0 \times [0,t_1].$$

It follows from (5.3.62) and the condition (iii) that

$$- \int_{B_0} (\tfrac{1}{T}\theta)_{,A}\Phi_A dV \leqq \tfrac{1}{2}(\alpha_1 + \alpha_2 - 2K_o)\int_{B_0} \theta_{,A}\theta_{,A}dV + \\ + c_o \int_{B_0} (\theta^2 + u_{i,A}u_{i,A})dV, \tag{5.3.63}$$

where

$$c_o = \tfrac{1}{2}\max(m_1^2\,\alpha_1^{-1} + 2m_4^2 + 1, m_2^2\,\alpha_2^{-1} + m_3^2).$$

If α_1 and α_2 are chosen such that

$$\alpha_1 + \alpha_2 - 2K_o \leqq 0,$$

then the inequality (5.3.63) implies the desired result. □

With a view toward deriving the continuous dependence result, we record the following Gronwall-type inequality (Dafermos [92]).

Lemma 5.3.8. Assume that the nonnegative functions $f \in L^{\infty}([0,t_1])$ and $g \in L^1([0,t_1])$ satisfy the inequality

$$f^2(t) \leqq M^2 f^2(0) + 2 \int_0^t [mf^2(\tau) + Ng(\tau)f(\tau)]d\tau, \quad t \in [0,t_1],$$

where m,M and N are nonnegative constants. Then

$$f(t) \leqq [Mf(0) + N \int_0^t g(s)ds]\exp(mt), \qquad t \in [0,t_1].$$

Let Y and Z be the functions on $[0,t_1]$ defined by

$$Y = \{\int_{B_0} (\dot{u}_i\dot{u}_i + u_{i,A}u_{i,A} + \theta^2)dV\}^{1/2}$$
$$Z = \{\int_{B_0} (F_iF_i + S^2)dV\}^{1/2}. \tag{5.3.64}$$

Theorem 5.3.3. Let $(\underline{u},\theta)$ be a solution of the problem (P). Then there exist positive constants m,M, and N such that

$$Y(t) \leqq [MY(0) + N \int_0^t Z(s)ds]\exp(mt), \quad t \in [0,t_1].$$

Proof. By (5.3.60) and (5.3.61), we obtain

$$\dot{y} \leqq \int_{B_0} \rho_0(F_i\dot{u}_i + \frac{1}{T} S\theta)dV + \frac{1}{2}\int_{B_0} (\dot{D}_{iAjK}u_{i,A}u_{j,K} - \dot{A}\theta^2 -$$
$$- 2\dot{F}_{Ki}u_{i,K}\theta - \frac{2}{T} \rho_0\dot{\eta}\theta^2)dV + c_o \int_{B_0} (u_{i,K}u_{i,K} + \theta^2)dV.$$

By using the inequality (5.3.11) and the Schwarz inequality, we arrive at

$$\dot{y} \leqq \nu Z[\int_{B_0} (\dot{u}_i\dot{u}_i + \theta^2)dV]^{1/2} + c_1 \int_{B_0} (u_{i,A}u_{i,A} + \theta^2)dV, \tag{5.3.65}$$

where ν and c_1 are positive constants. It follows from (5.3.65) and the hypotheses (i), (ii) that

$$Y^2(t) \leqq M^2Y^2(0) + 2 \int_0^t [mY^2(s) + NZ(s)Y(s)]ds, \tag{5.3.66}$$

where

$$M^2 = \frac{m_o}{c_2}, \quad m^2 = \frac{c_1}{2c_2}, \quad N = \frac{\nu}{2c_2}, \quad m_o = \frac{1}{2}\max(\rho_o, D_{iAjK}, A)$$

$$\text{on } \bar{B}_o \times [t_o, t_1],$$

$$c_2 = \frac{1}{2}\min(\rho_1, D_o, A_o).$$

The inequality (5.3.66) and Lemma 5.3.8 imply the desired result. □

Other types of boundary conditions may also be considered. The following proposition is a direct consequence of the above result.

Theorem 5.3.4. Assume that the hypotheses (i)-(iv) hold. Then the initial-boundary-value problem of thermoelasticity has at most one solution.

5.4 Reciprocal theorem. Variational theorem

Throughout this section we assume that the primary state is a configuration of equilibrium at constant temperature. We shall present a reciprocal theorem and a variational theorem. The reciprocal theorem for the classical theory of thermoelastodynamics is due to Ionescu-Cazimir [203] whose proof is based on the assumption of null initial data and systematic use of the Laplace transform. Later, Ieşan [186], [188] gave a general reciprocal theorem utilizing the properties of convolution. We now establish a reciprocal theorem in the context of the linear theory of thermoelastic bodies with initial stress [196].

Let $L^{(\alpha)} = (\underline{F}^{(\alpha)}, S^{(\alpha)}, \tilde{u}^{(\alpha)}, \underline{\tilde{p}}^{(\alpha)}, \tilde{\theta}^{(\alpha)}, \tilde{\phi}^{(\alpha)}, \underline{\tilde{a}}^{(\alpha)}, \underline{\tilde{b}}^{(\alpha)}, \tilde{\theta}_o^{(\alpha)})$ $(\alpha = 1,2)$ be two external data systems. Further, let $P^{(\alpha)} = (u_i^{(\alpha)}, \theta^{(\alpha)}, s_{jk}^{(\alpha)}, \phi_i^{(\alpha)})$ be the corresponding thermoelastic processes. We introduce the notations

$$\begin{aligned} h_j^{(\alpha)} &= \rho(i * F_j^{(\alpha)} + \tilde{a}_j^{(\alpha)} + t\tilde{b}_j^{(\alpha)}), \quad s_j^{(\alpha)} = s_{kj}^{(\alpha)} n_k, \\ g^{(\alpha)} &= 1 * \rho S^{(\alpha)} + aT\tilde{\theta}_o^{(\alpha)} + T\beta_{ij}\tilde{a}_{i,j}^{(\alpha)}, \quad q^{(\alpha)} = \phi_i^{(\alpha)} n_i. \end{aligned} \tag{5.4.1}$$

Theorem 5.4.1. Assume that the conductivity tensor k_{ij} is symmetric. Let $P^{(\alpha)}$ be thermoelastic processes corresponding to the external data systems $L^{(\alpha)}$ $(\alpha = 1,2)$. Then

$$\int_B [\underline{h}^{(1)} * \underline{u}^{(2)} - \frac{1}{T} i * g^{(1)} * \theta^{(2)}]dv + \int_{\partial B} i * (\underline{s}^{(1)} * \underline{u}^{(2)} - - \frac{1}{T} * q^{(1)} * \theta^{(2)})da = \int_B [\underline{h}^{(2)} * \underline{u}^{(1)} - \frac{1}{T} i * g^{(2)} * \theta^{(1)}]dv + + \int_{\partial B} i * (\underline{s}^{(2)} * \underline{u}^{(1)} - \frac{1}{T} * q^{(2)} * \theta^{(1)})da. \tag{5.4.2}$$

Proof. By (5.1.15) and (5.1.17),

$$s_{ji}^{(1)} * u_{i,j}^{(2)} + \frac{1}{J} \theta^{(1)} * \gamma^{(2)} = s_{ji}^{(2)} * u_{i,j}^{(1)} + \frac{1}{J} \theta^{(2)} * \gamma^{(1)}. \tag{5.4.3}$$

If we define

$$D_{\alpha\beta} = \int_B i * (s_{ji}^{(\alpha)} * u_{i,j}^{(\beta)} + \frac{1}{J} \theta^{(\alpha)} * \gamma^{(\beta)})dv, \tag{5.4.4}$$

then (5.4.3) implies

$$D_{12} = D_{21}. \tag{5.4.5}$$

By using (5.1.15) and (5.2.6),

$$i * (s_{ji}^{(\alpha)} * u_{i,j}^{(\beta)} + \frac{1}{J} \theta^{(\alpha)} * \gamma^{(\beta)}) = i * (s_{jk}^{(\alpha)} * u_k^{(\beta)} + + \frac{1}{T} * \theta^{(\alpha)} * \phi_j^{(\beta)})_{,j} + h_i^{(\alpha)} * u_i^{(\beta)} + \frac{1}{T} i * \theta^{(\alpha)} * g^{(\beta)} - - \rho u_i^{(\alpha)} * u_i^{(\beta)} - \frac{1}{T} i * 1 * k_{rs} \theta_{,s}^{(\beta)} * \theta_{,r}^{(\alpha)}.$$

In view of the divergence theorem we find that

$$D_{\alpha\beta} = \int_{\partial B} i * (\underline{s}^{(\alpha)} * \underline{u}^{(\beta)} + \frac{1}{T}\theta^{(\alpha)} * q^{(\beta)})da +$$

$$+ \int_B [\underline{h}^{(\alpha)} * \underline{u}^{(\beta)} + \frac{1}{T} i * \theta^{(\alpha)} * g^{(\beta)}]dv - \qquad (5.4.6)$$

$$- \int_B [\rho\underline{u}^{(\alpha)} * \underline{u}^{(\beta)} + \frac{1}{T} i * 1 * k_{rs}\theta^{(\beta)}_{,s} * \theta^{(\alpha)}_{,r}]dv.$$

From the symmetry of k_{ij} and (5.4.5) we obtain (5.4.2). □

It is possible to give applications of Theorem 5.4.1, similar to those given in the classical theory (see, for example [203], [47]).

We now derive a variational theorem. In the context of the classical theory of thermoelasticity, variational theorems have been established by Ieşan [185] and Nickell and Sackmann [272] (see also Carlson [47], Sect. 24, and Lebon [241], p. 370). A variational theorem for the linear theory of thermoelastic bodies with initial stress was established in [196].

We say that $v = (u_1,u_2,u_3,\theta) = (\underline{u},\theta)$ is an admissible four-dimensional vector field if $\underline{u}$ is an admissible displacement field on $\bar{B} \times [0,t_1)$ and θ is an admissible temperature field on $\bar{B} \times [0,t_1)$.

Let Q be the set of all admissible four-dimensional vector fields $v = (\underline{u},\theta)$ that satisfy the boundary conditions

$$\underline{u} = \tilde{\underline{u}} \text{ on } \bar{S}_1 \times [0,t_1), \quad \theta = \tilde{\theta} \text{ on } \bar{S}_3 \times [0,t_1). \qquad (5.4.7)$$

<u>Theorem 5.4.2.</u> Assume that the conductivity tensor k_{ij} is symmetric. Let $v = (\underline{u},\theta) \in Q$ and for each $t \in [0,t_1)$ define the functional $F_t\{\cdot\}$ on Q by

$$F_t\{v\} = \int_B \{i * [d_{ijrs}u_{r,s} * u_{i,j} - 2\beta_{ij}u_{i,j} * \theta - a\,\theta*\theta -$$

$$- \frac{1}{T}(2g * \theta + 1 * k_{ij}\theta_{,j} * \theta_{,i})] + \rho u_i * u_i - 2h_i * u_i\}dv -$$

$$- 2\int_{S_2} i * \tilde{p}_j * u_j da + \frac{2}{T}\int_{S_4} 1 * i * \tilde{q} * \theta\, da.$$

Then

$$\delta F_t\{v\} = 0 \quad (0 \leqq t < t_1), \tag{5.4.8}$$

at $v \in Q$ if and only if v is a solution of the mixed problem.

Proof. Let $v^o = (\underline{u}^o, \theta^o)$ be an admissible four-dimensional vector field, and suppose that

$$v + \lambda v^o \in Q \text{ for every scalar } \lambda. \tag{5.4.9}$$

This later condition is equivalent to the requirement that v^o meets the boundary conditions

$$\underline{u}^o = \underline{0} \text{ on } \bar{S}_1 \times [0,t_1), \quad \theta^o = 0 \text{ on } \bar{S}_3 \times [0,t_1). \tag{5.4.10}$$

By (5.1.17), (5.4.10), and the divergence theorem we arrive at

$$\begin{aligned} \delta_{v^o} F_t\{v\} = 2 \int_B &\{[\rho u_i - h_i - i * (d_{ijrs} u_{r,s})_{,j}] * u_i^o + \\ &+ i * [\tfrac{1}{T} g - \beta_{ij} u_{i,j} - a\theta + \\ &+ \tfrac{1}{T} * (k_{ij}\theta_{,j})_{,i}] * \theta^o\} dv + \\ &+ 2 \int_{S_2} i * [(d_{ijrs} u_{r,s} - \beta_{ij}\theta) n_j - \tilde{p}_i] * u_i^o da + \\ &+ \frac{2}{T} \int_{S_4} i * 1 * (\tilde{q} - k_{ij}\theta_{,j} n_i) * \theta^o da \quad (0 \leqq t < t_1), \end{aligned} \tag{5.4.11}$$

for every admissible four-dimensional vector field $v^o = (\underline{u}^o, \theta^o)$ that satisfies (5.4.10). If v is a solution of the mixed problem, then (5.4.11), because of Theorem 5.2.2, yields (5.4.8). On the other hand (5.4.8), (5.4.11), Lemmas 3.4.1 and 3.4.2, and Theorem 5.2.2 imply that $v = (\underline{u}, \theta)$ is a solution of the mixed problem. □

The form of the functional $F_t\{\cdot\}$ has been obtained by the method given in [190].

5.5. Existence theorem

In this section we apply the theory of semigroups of linear operators to establish an existence and uniqueness result for the boundary-initial-value problems of thermoelasticity. The results we give here are due to Navarro and Quintanilla [271].

We consider that the primary state is a time-dependent configuration. The behaviour of the body is studied on a finite time interval $[0,t_1]$. We consider the equations (5.1.1)-(5.1.3) on $B_o \times (0,t_1)$. To these equations we adjoin the initial conditions (5.2.8) and the boundary conditions

$$\underline{u} = \underline{0}, \quad \theta = 0 \text{ on } \partial B_o \times [0,t_1). \tag{5.5.1}$$

The field equations can be written in the form (5.2.10). We assume that $A \neq 0$ on $B_o \times [0,t_1]$. Then, the equations (5.2.10) may be written in the form

$$\begin{aligned} \ddot{u}_i &= \frac{1}{\rho_0}(D_{iAjK}u_{j,K} - F_{Ai}\theta)_{,A} + F_i, \\ \dot{\theta} &= \frac{1}{AT}[(K_{AM}\theta_{,M} + D_A\theta + G_{AiK}u_{i,K})_{,A} - \rho_o\dot{\eta}\theta] - \\ &- \frac{1}{A}(F_{Ai}\dot{u}_{i,A} + \dot{F}_{Ai}u_{i,A} + \dot{A}\theta) + \frac{\rho_0}{AT}S, \end{aligned} \tag{5.5.2}$$

on $B_o \times (0,t_1)$. We first transform the boundary-initial-value problem into a temporally inhomogeneous abstract equation in a Hilbert space. Then we use results of the semigroups theory of linear operators to derive an existence and uniqueness theorem. We assume that: (i) D_{iAjK}, G_{AiK}, F_{Ai}, K_{AM}, D_A, A, $\dot{A}$ and $\dot{F}_{Ai}$ are continuously differentiable in time on $[0,t_1]$, and Lebesgue measurable and essentially bounded on B_0, for each $t \in [0,t_1]$; (ii) $0 < \rho_1 \leqq \underset{X\in B_0}{\text{ess inf}}\, \rho_0 \leqq \underset{X\in B_0}{\text{ess sup}}\, \rho_0 \leqq \rho_2$, where ρ_1 and ρ_2 are constants; (iii) there exists a constant A_o such that $\underset{X\in B_0}{\text{ess inf}}\, A \geqq A_o > 0$ on $[0,t_1)$;

(iv) there exists a positive constant d such that

$$\int_{B_0} D_{iAjB}u_{i,A}u_{j,B}\,dV \geq d \int_{B_0} u_{i,K}u_{i,K}\,dV,$$

for all $u_i \in C_0^\infty(B_0)$ and any $t \in [0,t_1]$; (v) K_{MN} is positive definite in the sense that there exists a positive constant K_0 such that

$$\int_{B_0} \frac{1}{T} K_{AB}\theta_{,A}\theta_{,B}\,dV \geq K_0 \int_{B_0} \theta_{,A}\theta_{,A}\,dV,$$

for all $\theta \in C_0^\infty(B_0)$ and any $t \in [0,t_1]$.

Let $v_i = \dot{u}_i$ and define

$$X = \{w = (u_1,u_2,u_3,v_1,v_2,v_3,\theta) \equiv (u_i,v_i,\theta);\ u_i \in W_0^{1,2}(B_0),$$
$$v_i \in L_2(B_0),\ \theta \in L_2(B_0)\},$$

where $W_0^{1,2}(B_0)$ is the well-known Sobolev space [2]. Let

$$A_i w = v_i,\quad B_i w = \frac{1}{\rho_0}(D_{iAjK}u_{j,K} - F_{Ai}\theta)_{,A},$$

$$Cw = \frac{1}{AT}[(K_{LM}\theta_{,M} + D_L\theta + G_{LiK}u_{i,K})_{,L} - \rho_0\dot{\eta}\theta] - \tag{5.5.3}$$
$$- \frac{1}{A}(F_{Ki}v_{i,K} + \dot{F}_{Ki}u_{i,K} + \dot{A}\theta).$$

We introduce the operator Q defined by

$$Qw = (A_i w,\ B_i w,\ Cw), \tag{5.5.4}$$

with the domain

$$D = \{w = (u_i,v_i,\theta) \in X;\ Qw \in X,\ \theta = 0 \text{ on } \partial B_0 \times [0,t_1]\}. \tag{5.5.5}$$

The boundary-initial-value problem characterized by the relations (5.5.1), (5.5.2) and (5.2.8) can be reduced to the following abstract evolutionary equation

$$\frac{dw(t)}{dt} = Q(t)w(t) + G(t), \quad t \in [0,t_1],$$
$$w(0) = w_o, \tag{5.5.6}$$

where G and w_o are seven-dimensional vector fields defined by

$$G = (0,0,0,F_1,F_2,F_3,\rho_o S/(AT)), \quad w_o = (\hat{a}_i,\hat{b}_i,\hat{\theta}_o).$$

Let X_t be the Hilbert space X equipped with the norm $\|\cdot\|_t$, induced by the inner product

$$\langle w,\bar{w}\rangle_t = \frac{1}{2}\int_{B_o} [D_{iAjB}u_{i,A}\bar{u}_{j,B} + \rho_o v_i \bar{v}_i + A\theta\bar{\theta}]dV, \tag{5.5.7}$$

for each $t \in [0,t_1]$. By the hypotheses (ii)-(iv) and the Poincaré inequality we conclude that the norm $\|\cdot\|_t$ is equivalent to the original norm $\|\cdot\|$ in X.

Lemma 5.5.1. For each $t \in [0,t_1]$ the operator Q(t) is the generator of a quasi-contractive semigroup on X_t.

Proof. By (5.5.3), (5.5.4) and (5.5.7),

$$\langle Qw,w\rangle_t = \frac{1}{2}\int_{B_o} \{D_{iAjB}v_{i,A}u_{j,B} + v_i(D_{iAjK}u_{j,K} - F_{Ai}\theta)_{,A} +$$
$$+ \frac{1}{T}\theta[(K_{AM}\theta_{,M} + D_A\theta + G_{AiK}u_{i,K})_{,A} - \rho_o\dot{\eta}\theta -$$
$$- T(F_{Ki}v_{i,K} + \dot{F}_{Ki}u_{i,K} + \dot{A}\theta)]\}dV.$$

The divergence theorem and the boundary conditions imply

$$\langle Qw,w\rangle_t = -\frac{1}{2}\int_{B_o} \Phi_A\left(\frac{\theta}{T}\right)_{,A} dV - \frac{1}{2}\int_{B_o} \dot{F}_{Ai}u_{i,A}\theta\, dV -$$
$$-\frac{1}{2}\int_{B_o} (\rho_o \frac{1}{T}\dot{\eta} + \dot{A})\theta^2\, dV,$$

where Φ_A is given by (5.1.3). Using (5.3.11), Schwarz's inequality and (5.3.61) we conclude that there exists a positive constant β such that

$$\langle Qw,w\rangle_t \leqq \beta\langle w,w\rangle_t. \tag{5.5.8}$$

Our next objective is to show that the operator Q satisfies the range condition $R(\lambda I - Q) = X$. Let $\bar{w} = (\bar{u}_i,\bar{v}_i,\bar{\theta}) \in X$. We must prove that the equation

$$\lambda w - Qw = \bar{w}, \quad (\lambda > 0), \tag{5.5.9}$$

has a solution $w = (u_i,v_i,\theta) \in D$. By eliminating v_i, (5.5.9) yields the following system for u_i and θ

$$\begin{aligned} J_i z &\equiv \lambda^2 u_i - \frac{1}{\rho_0}(D_{iAjK}u_{j,K} - F_{Ai}\theta)_{,A} = m_i, \\ Jz &\equiv \lambda\theta - \frac{1}{AT}[(K_{LM}\theta_{,M} + D_L\theta + G_{LiK}u_{i,K})_{,L} - \rho_o\dot{\eta}\theta] - \\ &\quad - \frac{1}{A}(\lambda F_{Ki}u_{i,K} + \dot{F}_{Ki}u_{i,K} + \dot{A}\theta) = m, \end{aligned} \tag{5.5.10}$$

where

$$z = (u_1,u_2,u_3,\theta), \quad m_i = \bar{v}_i + \lambda\bar{u}_i, \; m = \bar{\theta} - \frac{1}{A}F_{Ki}\bar{u}_{i,K}. \tag{5.5.11}$$

Let $[\cdot,\cdot]$ denote a conveniently weighted L_2 inner product, and consider the bilinear form

$$f(z,\bar{z}) = [(J_i z,Jz),(\bar{u}_i,\bar{\theta})] = \int_{B_0}(k\rho_0\bar{u}_iJ_iz + A\bar{\theta}\,Jz)dV, \tag{5.5.12}$$

where k is an arbitrary positive constant. The divergence theorem and the boundary conditions imply

$$\begin{aligned} f(z,z) = \int_B [&k(\rho_0 u_i u_i + D_{iAjK}u_{i,A}u_{j,K} - F_{Ai}u_{i,A}\theta) + (K_{AM}\,\theta_{,M} + \\ &+ D_A\theta + G_{AiK}u_{i,A})(\tfrac{\theta}{T})_{,A} - \lambda F_{Ai}u_{i,A}\theta - \dot{F}_{Ai}u_{i,A}\theta + \end{aligned}$$

$$+ (A\lambda + \frac{1}{T}\rho_0\dot{\eta} - \dot{A})\theta^2]dV.$$

By use of the Schwarz and Young inequalities, and choosing conveniently the weighting constant k, it is not difficult to show that the bilinear form $f(z,\bar{z})$ determines a norm equivalent to

$$\|z\| = \|\underline{u}\|_{[W_0^{1,2}]^3} + \|\theta\|_{W_0^{1,2}}.$$

Next, we can deduce that $(\underline{m},m) \in [W^{-1,2}]^3 \times W^{-1,2}$ (see [271]). The Riesz theorem shows that there exists a unique pair $(\underline{u},\theta) \in [W_0^{1,2}]^3 \times W_0^{1,2}$. From (5.5.9) we obtain $v_i = \lambda u_i - \bar{u}_i$ so that $v_i \in \bar{W}_0^{1,2}$. We conclude that $w = (u_i,v_i,\theta) \in D$. The desired result is an immediate consequence of the Lumer-Phillips corollary to the Hille-Yosida theorem [385]. □

Lemma 5.5.2. There exists a positive constant α such that

$$\|w\|_t \leq e^{\alpha|t-s|} \|w\|_s, \tag{5.5.13}$$

holds for any $w = (u_i,v_i,\theta) \in X$ and each $s,t \in [0,t_1]$.

Proof. By (5.3.59) and (5.5.7), $y(t) = \|w\|_t^2$. It follows from (5.3.65) that there exists a positive constant α such that

$$|\dot{y}| \leq 2\alpha y.$$

Thus

$$\dot{y} \leq 2\alpha y. \tag{5.5.14}$$

The desired result follows from (5.5.14). □

The preceding lemmas imply the following result.

Lemma 5.5.3. The family of operators $\{Q(t)\}$, $t \in [0,t_1]$, is stable in the sense of Kato, with stability constants $\exp(2\alpha t_1)$ and β.

We now consider the problem (5.5.6) where w_0 and G are given elements of a Banach space Y, continuously and densely embedded in X.

The above lemmas and the Kato theorem [218] lead to the following proposition.

Theorem 5.5.1. Assume that the hypotheses (i)-(v) hold. Let F_i, $S \in C^1([0,t_1]; L_2(B_0))$. Then, for any $w_0 \in D$, there exists a unique solution to the boundary-initial-value problem (5.5.1), (5.5.2), (5.2.8) such that $(u_i,\dot{u}_i,\theta) \in C^1([0,t_1];X) \cap C([0,t_1];D)$.

5.6. Thermoelastic waves

α) Initially isotropic bodies. We assume that the body B_0 is homogeneous and isotropic. Further, we suppose that the primary state is obtained from B_0 by the isothermal uniform finite extensions

$$x_1 = \lambda_1 X_1, \; x_2 = \lambda_1 X_2, \quad x_3 = \lambda_3 X_3, \quad T = T_0, \tag{5.6.1}$$

where λ_i and T_0 are given constants. The special expressions assumed by the formulae of Sections 4.2 and 4.3 can be obtained without difficulty and results corresponding to $\theta = 0$ have already been given in Section 2.4. Consequently only the final formulae will be recorded here. It follows from (2.4.2), (4.2.20), (4.2.23), (4.2.32) - (4.2.34), (4.3.6) and (4.3.13) that

$$
\begin{aligned}
s_{11} &= a_{11}u_{1,1} + a_{12}u_{2,2} + a_{13}u_{3,3} - \beta_1\theta,\\
s_{22} &= a_{12}u_{1,1} + a_{11}u_{2,2} + a_{13}u_{3,3} - \beta_1\theta,\\
s_{33} &= a_{13}(u_{1,1} + u_{2,2}) + a_{33}u_{3,3} - \beta_3\theta,\\
s_{23} &= a_{44}u_{2,3} + a_{45}u_{3,2}, \quad s_{32} = a_{54}u_{2,3} + a_{44}u_{3,2},\\
s_{31} &= a_{54}u_{1,3} + a_{44}u_{3,1}, \quad s_{13} = a_{44}u_{1,3} + a_{45}u_{3,1},\\
s_{12} &= a_{66}u_{1,2} + a_{65}u_{2,1}, \quad s_{21} = a_{65}u_{1,2} + a_{66}u_{2,1},
\end{aligned} \tag{5.6.2}
$$

$$\gamma = J(\beta_1 u_{\alpha,\alpha} + \beta_3 u_{3,3} + a\theta),$$

$$\phi_1 = k_1\theta_{,1}, \ \phi_2 = k_1\theta_{,2}, \ \phi_3 = k_3\theta_{,3},$$

where a_{rs} $(r,s = 1,2,\ldots,6)$ are given by (2.4.12), (2.4.7), (2.4.8), with Φ, Ψ, p replaced by $\hat{\Phi}$, $\hat{\Psi}$, $\hat{p}$, respectively, and

$$\begin{aligned}
\beta_1 &= -\lambda_1^2 L_1 - \lambda_1^2(\lambda_1^2 + \lambda_3^2)L_2 - \lambda_1^4\lambda_3^2 L_3,\\
\beta_3 &= -\lambda_3^2 L_1 - 2\lambda_1^2\lambda_3^2 L_2 - \lambda_1^4\lambda_3^2 L_3,\\
k_1 &= \frac{1}{\lambda_3}[K_1 + \tfrac{1}{2}K_2(\lambda_1^2 - 1) + \tfrac{1}{4}K_3(\lambda_1^2 - 1)],\\
k_3 &= \frac{\lambda_3}{\lambda_1^2}[K_1 + \tfrac{1}{2}K_2(\lambda_3^2 - 1) + \tfrac{1}{4}K_3(\lambda_3^2 - 1)],\\
t_{11} &= t_{22} = \lambda_1^2\hat{\Phi} + \lambda_1^2(\lambda_1^2 + \lambda_3^2)\hat{\Psi} + \hat{p},\\
t_{33} &= \lambda_3^2\hat{\Phi} + 2\lambda_1^2\lambda_3^2\hat{\Psi} + \hat{p}, \quad t_{ij} = 0 \ (i \neq j).
\end{aligned} \tag{5.6.3}$$

Clearly, the coefficients a_{rs} $(r,s = 1,2,\ldots,6)$, β_i, a and k_i are constants. Substitution of (5.6.2) into the equations of motion and the energy equation yields the equations

$$\begin{aligned}
&a_{11}u_{1,11} + \tfrac{1}{2}(a_{11}-d_{12})u_{1,22} + a_{54}u_{1,33} + \tfrac{1}{2}(a_{11}+d_{12})u_{2,12} +\\
&\qquad + (d_{13}+a_{45})u_{3,13} - \beta_1\theta_{,1} + \rho F_1 = \rho\ddot{u}_1,\\
&\tfrac{1}{2}(a_{11}+d_{12})u_{1,12} + \tfrac{1}{2}(a_{11}-d_{12})u_{2,11} + a_{11}u_{2,22} +\\
&\qquad + a_{54}u_{2,33} + (d_{13}+a_{45})u_{3,23} - \beta_1\theta_{,2} +\\
&\qquad + \rho F_2 = \rho\ddot{u}_2,
\end{aligned} \tag{5.6.4}$$

$$(d_{13}+a_{45})u_{\alpha,\alpha 3} + a_{45}\Delta_o u_3 + a_{33}u_{3,33} - \beta_3\theta_{,3} +$$

$$+ \rho F_3 = \rho \ddot{u}_3,$$

$$k_1 \Delta_o \theta + k_3 \theta_{,33} - aT_o \dot{\theta} - T_o(\beta_1 \dot{u}_{\alpha,\alpha} + \beta_3 \dot{u}_{3,3}) = - \rho S,$$

where d_{12} and d_{13} are given by (2.4.14), and $\rho \lambda_1^2 \lambda_3 = \rho_0$.

It follows from (4.2.36) that

$$k_1 \geqq 0, \quad k_3 \geqq 0. \tag{5.6.5}$$

a) <u>Plane waves</u>. We now apply the equations (5.6.4) to the study of plane harmonic waves. The results we give here are due to Flavin and Green [122]. These results generalize those given by Chadwick [52] in the classical theory of thermoelasticity.

Following [122], we define

$$e = u_{\alpha,\alpha}, \quad W = u_{1,2} - u_{2,1}, \; f = u_{3,3}. \tag{5.6.6}$$

We assume that $F_i = 0$ and $S = 0$. The equations (5.6.4) imply

$$\frac{1}{2}(a_{11} - d_{12}) \Delta_o W + a_{54} W_{,33} = \rho \ddot{W}, \tag{5.6.7}$$

$$a_{11} \Delta_o e + a_{54} e_{,33} + (d_{13} + a_{45}) \Delta_o f - \beta_1 \Delta_o \theta = \rho \ddot{e},$$

$$a_{45} \Delta_o f + a_{33} f_{,33} - \rho \ddot{f} + (d_{13} + a_{45}) e_{,33} - \beta_3 \theta_{,33} = 0, \tag{5.6.8}$$

$$k_1 \Delta_o \theta + k_3 \theta_{,33} - m \dot{\theta} - T_0(\beta_1 \dot{e} + \beta_3 \dot{f}) = 0,$$

where $m = aT_o$. From (2.4.8), (2.4.12) and (5.6.3), we obtain

$$a_{11} - d_{12} = 2\lambda_1^2(\hat{\Phi} + \lambda_3^2 \hat{\Psi}), \quad a_{54} = \lambda_3^2(\hat{\Phi} + \lambda_1^2 \hat{\Psi}). \tag{5.6.9}$$

We seek W of the form

$$W = \mathrm{Re}\{w_o \exp[i(\eta \ell_j x_j - \omega t)]\}, \tag{5.6.10}$$

which represents a plane wave. Here $\underline{\ell}$ is a unit vector, η is the wave number, and w_o a constant amplitude. We define wave velocity by $v_2 = \omega/\eta$. It follows from (5.6.7) and (5.6.10) that

$$v_2^2 = \lambda_1^2\lambda_3\{[\lambda_1^2(1 - \ell_3^2) + \lambda_3^2\ell_3^2]\hat{\Phi} + \lambda_1^2\lambda_2^2\hat{\Psi}\}/\rho_0. \qquad (5.6.11)$$

Clearly, (5.6.10) represents a transverse wave, with vibrations in planes parallel to the plane $x_1 0 x_2$. This wave is independent of thermal effects. If the initial deformation is zero, then v_2 reduces to $(\mu/\rho_0)^{1/2}$, where μ is the second Lamé constant. We now consider the equations (5.6.8). Let

$$u_\alpha = (f_1 + f_2)_{,\alpha}, \quad u_3 = (h_1 f_1 + h_2 f_2)_{,3}, \qquad (5.6.12)$$

where f_α are class C^2 fields on $B \times (0,t_1)$, and h_α are given by (2.5.1), (2.5.2). It follows from (5.6.12) that

$$W = 0, \quad e = \Delta_o(f_1+f_2), \quad f = (h_1 f_1 + h_2 f_2)_{,33}.$$

The equations (5.6.8) are satisfied if

$$\begin{aligned} & a_{11}(\Delta_o f_1 + m_1 f_{1,33}) - \rho\ddot{f}_1 + a_{11}(\Delta_o f_2 + m_2 f_{2,33}) - \\ & \qquad - \rho\ddot{f}_2 - \beta_1\theta = 0, \\ & r_2(\Delta_o f_1 + m_1 f_{1,33}) - \rho h_1\ddot{f}_1 + r_1(\Delta_o f_2 + m_2 f_{2,33}) - \\ & \qquad - \rho h_2\ddot{f}_2 - \beta_3\theta = 0, \\ & k_1(\Delta_o\theta + b\theta_{,33}) - m\dot{\theta} - T_0(\beta_1\Delta_o\dot{f}_1 + \beta_3 h_1\dot{f}_{1,33} + \\ & \qquad + \beta_1\Delta_o\dot{f}_2 + \beta_3 h_2\dot{f}_{2,33}) = 0, \end{aligned} \qquad (5.6.13)$$

where

$$b = k_3/k_1, \quad r_\alpha = m_\alpha a_{11}/h_\alpha. \qquad (5.6.14)$$

We seek solutions to (5.6.13) of the type

$$(f_1, f_2, \theta) = \text{Re}\,\{(a_1, a_2, a_3)\exp[i(\eta \ell_s x_s - \omega t)]\}, \quad \underline{\ell}^2 = 1. \tag{5.6.15}$$

If we put

$$A_\alpha = 1 - \ell_3^2 + m_\alpha \ell_3^2, \quad A_3 = 1 - \ell_3^2 + b\ell_3^2, \quad \rho v_1^2 = a_{11},$$

$$k_1\omega^* = mv_1^2, \ \chi = \omega/\omega^*, \quad \xi = v_1\eta/\omega^*, \quad g_i = -\beta_i/m \ (i = 1,3), \tag{5.6.16}$$

$$h = mT_o/a_{11},$$

the condition for plane waves is

$$\det(A_{ij}) = 0, \tag{5.6.17}$$

where

$$A_{11} = A_1\xi^2 - \chi^2, \quad A_{12} = A_2\xi^2 - \chi^2, \quad A_{13} = -g_1,$$

$$A_{23} = -g_2, \quad A_{21} = \frac{m_2}{h_2} A_1\xi^2 - h_1\chi^2,$$

$$A_{22} = \frac{m_1}{h_1} A_1\xi^2 - h_2\chi^2, \quad A_{3\alpha} = -ih\xi^2\chi[g_1(1-\ell_3^2) + g_3h_\alpha\ell_3^2], \tag{5.6.18}$$

$$A_{33} = A_3\xi^2 - i\chi.$$

We examine the roots of the equation (5.6.17) when $\chi \ll 1$. It follows that one root of (5.6.17) is given by

$$\xi^2 = -iA_3^{-1}(1+s)\chi + O(\chi^2), \tag{5.6.19}$$

where

$$s = h(A_1A_2)^{-1}\{\ell_3^2(1-\ell_3^2)(g_3-m_1g_1/h_1)(g_3-m_2g_2/h_2)a_{11}/a_{55} +$$

$$+ [g_1(1-\ell_3^2) + \ell_3^2h_1g_3][g_1(1-\ell_3^2) + \ell_3^2h_2g_3]\}.$$

The remaining roots are given by

$$\xi^2 = z\theta^2 + O(\chi^3), \tag{5.6.20}$$

where z is defined by the equation

$$(1+s)A_1A_2z^2 - \{1-\ell_3^2(1-h_1h_2) + (a_{11}/a_{45})[(1 + hg_1^2)(1 - \ell_3^2) + \\ + (hg_3^2 + m_1m_2/h_1h_2)\ell_3^2]\}z + a_{11}/a_{45} = 0. \tag{5.6.21}$$

In the limiting case of no initial deformation the value of ξ^2 given by (5.6.19) reduces to the quasi-thermal mode of classical theory of thermoelasticity (see Chadwick [52]). Moreover, the roots of the equation (5.6.21) become $z_1 = 1/(1 + hg_1^2)$ and $z_2 = (\lambda + 2\mu)/\mu$. The wave corresponding to (5.6.20), with $z = z_1$, reduces to the longitudinal wave of the classical thermoelasticity, and the wave corresponding to (5.6.20), with $z = z_2$, reduces to the ordinary transverse wave. We now discuss two special cases in which the equation (5.6.17) simplifies.

i) <u>Plane waves propagating in the direction of the x_3-axis.</u> In this case $\ell_3 = 1$, and the equation (5.6.17) becomes

$$(\xi^2a_{54}-\chi^2a_{11})[(b\xi^2-i\chi)(\xi^2a_{33}-\chi^2a_{11})-i\xi^2g_3^2ha_{11}\chi] = 0. \tag{5.6.22}$$

The solution

$$\xi^2 = \chi^2a_{11}/a_{54}, \tag{5.6.23}$$

corresponds to a transverse wave propagated with velocity

$$(a_{54}/\rho)^{1/2} = \lambda_1\lambda_3[\lambda_3(\hat{\Phi} + \lambda_1^2\hat{\Psi})/\rho_0]^{1/2}. \tag{5.6.24}$$

This velocity coincides with that given by (5.6.11) when $\ell_3 = 1$. The remaining roots of (5.6.22) are

$$\xi^2 = \chi^2(hg_3^2 + a_{33}/a_{11})^{-1} + O(\chi^3),$$

and

$$\xi^2 = i(1 + ha_{11}g_3^2/a_{33})\chi/b + O(\chi^2).$$

ii) Plane waves propagating in planes parallel to the $x_1 0 x_2$-plane. In this case $\ell_3 = 0$ and the equation (5.6.17) has a solution given by

$$\xi^2 = \chi^2 a_{11}/a_{45}. \qquad (5.6.25)$$

The corresponding wave velocity is

$$(a_{45}/\rho)^{1/2} = [(\lambda_1^2\hat{\Phi} + \lambda_1^4\hat{\psi})/\rho]^{1/2}. \qquad (5.6.26)$$

We note that the transverse wave velocity given by (5.6.11) is, when $\ell_3 = 0$, equal to

$$\lambda_1^2[\lambda_3(\hat{\Phi} + \lambda_3^2\hat{\Psi})/\rho_0]^{1/2}.$$

The remaining roots of (5.6.17) when $\ell_3 = 0$, are

$$\xi^2 = \chi^2(1 + hg_1^2)^{-1} + O(\chi^3), \; \xi^2 = i(1 + hg_1^2)\chi + O(\chi^2).$$

b) Rayleigh waves. We now study the propagation of thermoelastic Rayleigh waves in a half space which has been subjected to large uniform extensions, at constant temperature, given by (5.6.1). The result we give here is due to Flavin [123]. Thermoelastic Rayleigh waves within the classical theory have been studied by Chadwick [52].

We assume that the body B occupies the half space $x_3 \geqq 0$, and that the stress component t_{33} is zero. We then consider a thermoelastic deformation of B such that

$$u_1 = u_1(x_1,x_3,t),\; u_2 = 0,\; u_3 = u_3(x_1,x_3,t), \theta = \theta(x_1,x_3,t). \qquad (5.6.27)$$

We assume that the body forces and the heat supply are absent. Substitution of (5.6.27) into the equations (5.6.4) yields the equations

$$a_{11}u_{1,11} + a_{54}u_{1,33} + (d_{13}+a_{45})u_{3,13} - \beta_1\theta_{,1} = \rho\ddot{u}_1,$$

$$(d_{13}+a_{45})u_{1,13} + a_{45}u_{3,11} + a_{33}u_{3,33} - \beta_3\theta_{,3} = \rho\ddot{u}_3, \qquad (5.6.28)$$

$$k_1\theta_{,11} + k_3\theta_{,33} - m\dot{\theta} - T_0(\beta_1\dot{u}_{1,1} + \beta_3\dot{u}_{3,3}) = 0,$$

on $B \times (0,t_1)$. We assume that the boundary $x_3 = 0$ is a free surface so that

$$s_{3i} = 0 \text{ on } x_3 = 0. \qquad (5.6.29)$$

In addition we have the thermal boundary condition

$$\theta_{,3} + h_o\theta = 0 \text{ on } x_3 = 0, \qquad (5.6.30)$$

where h_o is a given constant. This condition corresponds to a thermally radiating boundary.

We seek solutions of the form

$$(u_1,u_3,\theta) = \text{Re}[(u_1^0,u_3^0,\theta^0)\exp\{[i(x_1-ct)-qx_3]\omega/c\}]. \qquad (5.6.31)$$

Clearly, (5.6.31) represents a Rayleigh wave of frequency $\omega/(2\pi)$. The physical reasoning requires that

$$\text{Re}(q/c) > 0. \qquad (5.6.32)$$

Since $t_{33} = 0$, we conclude from (5.6.2) and (5.6.27) that the boundary conditions (5.6.29) reduce to

$$a_{13}u_{1,1} + a_{33}u_{3,3} = \beta_1\theta, \quad u_{1,1} + u_{3,3} = 0 \text{ on } x_3 = 0. \qquad (5.6.33)$$

It follows from (5.6.28) and (5.6.31) that

$$u_1 = \text{Re}\{i(v_1/\omega^*) \sum_{s=1}^{3} [g_3b_4 - g_1b_1)q_s^2 - g_1(y^2-b_2)]A_sE_s\},$$

$$u_3 = \text{Re}\{i(v_1/\omega^*) \sum_{s=1}^{3} [g_3(b_2 q_s^2 - 1 + y^2) + g_1 b_4] q_s A_s E_s\}, \tag{5.6.34}$$

$$\theta = \text{Re}\{(T_0/hy)\chi \sum_{s=1}^{3} A_s L(q_s) E_s\},$$

where we have used the notations (5.6.14), (5.6.16), and

$$\begin{aligned} E_s &= \exp\{[i(x_1 - ct) - q_s x_3]\omega/c\}, \\ L(q) &= (b_2 q^2 - 1 + y^2)(b_1 q^2 - b_3 + y^2) + b_4^2 q^2, \\ b_1 &= a_{33}/a_{11},\ b_2 = a_{54}/a_{11},\ b_3 = a_{45}/a_{11}, \\ b_4 &= (d_{13} + a_{45})/a_{11},\ y = c/v_1. \end{aligned} \tag{5.6.35}$$

In the relations (5.6.34), A_s are arbitrary constants and q_j are the roots of the equation

$$\begin{aligned} &[b_2 q^2 - (1 + hg_1^2) + y^2][(b_1 + hg_3^2)q^2 - b_3 + y^2]y^2 + \\ &\quad + (b_4 + hg_1 g_3)^2 q^2 y^2 - i\chi(bq^2 - 1)L(q) = 0. \end{aligned} \tag{5.6.36}$$

In view of (5.6.30), (5.6.33) and (5.6.34), the boundary conditions on $x_3 = 0$ reduce to

$$p_{ij} A_j = 0, \tag{5.6.37}$$

where

$$\begin{aligned} p_{1s} &= b_2 q_s^2 [g_3(y^2 - b_3 + b_4) - g_1 b_1] + (y^2 - b_3)[g_3(y^2 - 1) + \\ &\quad + g_1(b_4 - b_2)], \\ p_{2s} &= [g_3(b_4 - b_2) - g_1 b_1] q_s^3 - [g_1(y^2 - b_3 + b_4) + g_3(y^2 - 1)] q_s, \\ p_{3s} &= [1 - \chi q_s/(jy)] L(q_s), \qquad j = h_o v_1/\omega^*. \end{aligned} \tag{5.6.38}$$

For non-zero solutions of (5.6.37) we have

$$\det(p_{ij}) = 0. \tag{5.6.39}$$

The equations (5.6.36) and (5.6.39) determine y and q_j in terms of χ. We now suppose that $\chi \ll 1$ and take $y^2 = O(1)$ as $\chi \to 0$. Then $q_3^2 = O(\chi^{-1})$, and the remaining roots of the equations (5.6.36) satisfy

$$\begin{aligned} b_2(b_1 + hg_3^2)(q_1^2 + q_2^2) &= -(b_4 + hg_1g_3)^2 + \\ &\quad + (b_1 + hg_3^2)(1 + hg_1^2 - y^2) + b_2(b_3 - y^2), \\ b_2(b_1 + hg_3^2)q_1^2q_2^2 &= (1 + hg_1^2 - y^2)(b_3 - y^2), \end{aligned} \tag{5.6.40}$$

where we have omitted the terms of order $O(\chi)$. Using (5.6.38) and (5.6.40) we can write the equation (5.6.39) in the form

$$(q_1 - q_2)G(y)[F(y) + O(\chi^{1/2})] = 0, \tag{5.6.41}$$

where

$$\begin{aligned} F(y) = (y^2-b_3+b_2)(b_1+hg_3^2)b_2q_1q_2 - [(1+hg_1^2-y^2)(b_1+hg_3^2) - \\ - (b_4+hg_1g_3-b_2)^2](b_3-y)^2, \end{aligned}$$

$$G(y) = g_3[g_3b_4-g_1(b_1-b_2)]y^2 + (g_3-g_1b_4)(g_1b_1-g_3b_4)-g_1g_3b_2b_3.$$

If $q_1 = q_2$ or $G(y) = 0$, then $u_1 = u_3 = 0$ and $\theta = 0$. In view of (5.6.40), the equation (5.6.41) reduces to

$$\begin{aligned} (b_3-y^2)[(1+hg_1^2-y^2)(b_1+hg_3^2)-(b_4+hg_1g_3-b_2)^2]^2 - \\ -b_2(b_1+hg_3^2)(1+hg_1^2-y^2-b_3+b_4)^2 + O(\chi^{1/2}) = 0. \end{aligned} \tag{5.6.42}$$

If we consider wave motion under isentropic or isothermal conditions, then it can be shown that the values of y^2 differ from the corresponding values derived from the equation (5.6.42) by terms of order $O(\chi^{1/2})$ only.

β) Anisotropic bodies. We now assume that the body B_0 is homogeneous and that the primary state is obtained from B_0 by the isothermal homogeneous deformation

$$x_i = \lambda_{iK}X_K, \quad T = T_0,$$

where λ_{iK} and T_0 are given constants. We assume that the body forces and heat supplies are absent. Then, the fundamental system (5.2.3) of field equations describing the infinitesimal thermoelastic deformations becomes

$$d_{ijrs}u_{r,sj} - \beta_{ij}\theta_{,j} = \rho\ddot{u}_i, \quad k_{ij}\theta_{,ij} - T_0\beta_{ij}\dot{u}_{i,j} = aT_o\dot{\theta}, \tag{5.6.43}$$

where d_{ijrs}, β_{ij}, k_{ij}, a and ρ are given constants. We assume that: (i) d_{ijrs} is strongly elliptic; (ii) k_{ij} is positive definite; (iii) a is positive; (iv) (β_{ij}) is nonsingular. We apply the equations (5.6.43) to the study of plane harmonic waves. The results we give here are due to Chadwick [60].

We seek solutions to (5.6.43) of the form

$$(u_r, \theta) = \text{Re}\,\{(U_r, \tau)\exp[i(\eta\ell_s x_s - \omega t)]\}, \tag{5.6.44}$$

where ω is supposed real and positive, and $\underline{\ell}^2 = 1$. It follows from (5.6.43) that the amplitudes U_j and τ satisfy the equations

$$\begin{aligned} &(G_{jr} - \rho s^{-2}\delta_{jr})U_r + i(\omega s)^{-1}\beta_{jm}\ell_m\tau = 0, \\ &T_0 s^{-1}\beta_{mn}U_m\ell_n + (K - aT_o i\omega^{-1}s^{-2})\tau = 0, \end{aligned} \tag{5.6.45}$$

where

$$G_{jr} = d_{jmrp}\ell_m\ell_p, \quad s = \eta/\omega, \quad K = k_{ij}\ell_i\ell_j. \tag{5.6.46}$$

The tensor G_{ij} is called the isothermal acoustic tensor for the direction $\underline{\ell}$. It follows from (1.2.45) and the hypothesis (i) that G_{ij} is symmetric and positive definite. The equations (5.6.45) have non-zero solutions if and only if

$$s^{-2}\det(D_{mn}-\rho s^{-2}\delta_{mn})+i\omega\,(aT_o)^{-1}K\,\det(G_{pq}-\rho s^{-2}\delta_{pq}) = 0, \qquad (5.6.47)$$

where

$$D_{ij} = G_{ij} + \frac{1}{a}\,\beta_{ir}\,\beta_{js}\,\ell_r\,\ell_s. \qquad (5.6.48)$$

The tensor D_{ij} is called the isentropic acoustical tensor associated with the direction $\underline{\ell}$. Clearly, D_{ij} is symmetric and positive definite.

The equation (5.6.47) determines s when the frequency ω, the direction $\underline{\ell}$, the finite homogeneous deformation and the constitutive coefficients are prescribed. The system (5.6.45) can be written in the form

$$\begin{aligned} &(i\omega K + aT_o s^{-2})\tau = -i\omega s^{-1}T_o\beta_{mn}U_m\ell_n, \\ &[\rho s^{-2}(D_{mp}-\rho s^{-2}\delta_{mp}) + i\omega\rho(aT_o)^{-1}K(G_{mp}-\rho s^{-2}\delta_{mp})]U_p = 0. \end{aligned} \qquad (5.6.49)$$

We introduce the notations

$$\begin{aligned} &w = \rho/ds^2,\ \omega^o = daT_o/\rho K,\ \zeta = i\omega/\omega^o = i\chi, \\ &d = G_{ij}\beta_{ir}\beta_{js}\ell_r\ell_s/\beta_{mp}\beta_{mq}\ell_p\ell_q. \end{aligned} \qquad (5.6.50)$$

Since G_{ij} is positive definite and β_{rs} is nonsingular, the scalar d is positive. We denote by κ_i and γ_p the proper numbers of the symmetric tensors $d^{-1}G_{ij}$ and $d^{-1}D_{rs}$, respectively. Since G_{rs} and D_{ij} are positive definite tensors, the constants κ_i and γ_j are positive. For the remainder of this section we assume that G_{mn} and D_{ij} both have distinct proper numbers. The equation (5.6.47) can be written in the form

$$w(w - \gamma_1)(w - \gamma_2)(w - \gamma_3) + \zeta(w - \kappa_1)(w - \kappa_2)(w - \kappa_3) = 0. \qquad (5.6.51)$$

This equation is a quartic equation in w with coefficients that are functions of ζ. The discriminant P of the polynomial in w forming the left side of the equation (5.6.51) is a sextic polynomial in ζ. Let Ω be the domain obtained by deleting from the ζ plane the zeros $\zeta_1,\zeta_2,\ldots,\zeta_6$ of the polynomial P. The equation (5.6.51) defines a four-valued algebraic function M, called the modal function, defined on Ω. The points $\zeta_1,\zeta_2,\ldots,\zeta_6$ are called singular points of M. The regular branches of M are denoted by w_i $(i = 1,2,3,4)$.

In view of (5.6.50), the equations (5.6.49) can be written in the form

$$d^{1/2} \rho^{-1/2}(w + \zeta)\tau = -\omega^o w^{1/2}\zeta\beta_{rs}\ell_s U_r, \tag{5.6.52}$$

$$[w(d^{-1}D_{ij} - w\delta_{ij}) + \zeta(d^{-1}G_{ij} - w\delta_{ij})]U_j = 0, \tag{5.6.53}$$

respectively. For a given frequency ω such that $\zeta \in \Omega$, the equations (5.6.51)-(5.6.53) and (5.6.44) associate with each regular branch of M a displacement vector and a temperature field. Thus, w_i $(i = 1,2,3,4)$ correspond to possible modes of waves propagation in the direction defined by $\underline{\ell}$. We now classify the branches of M with reference to the low-frequency limit $\zeta \to 0$. A classification based upon the high-frequency limit $\zeta \to \infty$ is presented in [60].

Clearly, the equation (5.6.51) for $\zeta = 0$ has the roots γ_i and 0. Since these roots are distinct, the point $\zeta = 0$ is a point of Ω. The branches of M can be labelled so that

$$w_i(0) = \gamma_i, \quad w_4(0) = 0. \tag{5.6.54}$$

Following [60], the modes corresponding to the branches w_1, w_2 and w_3 are called the quasielastic modes. The mode represented by w_4 is termed the quasithermal mode. Each mode can exist independently of the other three, and the corresponding displacement and temperature amplitudes are generally nonzero. A mode for which

$$\beta_{ij}\,\ell_j\,U_i = 0,$$

produces no change of temperature.

It is easy to see that the low-frequency limit, in which the wavelike modes are quasielastic modes, is identifiable with the theory of infinitesimal plane waves, in an elastic material that does not conduct heat.

The functions w_i (i = 1,2,3,4) admit Taylor expansions about the ordinary point $\zeta = 0$, which converge in the open disk centered at $\zeta = 0$ with the radius equal to the distance from the center to the nearest singular point of M (cf. Bliss [37]). Thus, for sufficiently small $|\zeta|$, we have

$$w_i = \gamma_i[1 + \sum_{n=1}^{\infty} c_n^{(i)}(-z)^n], \quad w_4 = \sum_{n=1}^{\infty} e_n(-z)^n. \qquad (5.6.55)$$

It follows from (5.6.51) and (5.6.55) that

$$c_1^{(i)} = g(\gamma_i)/[\gamma_i f'(\gamma_i)],$$

$$c_2^{(i)} = c_1^{(i)}[g'(\gamma_i) - \tfrac{1}{2} c_1^{(i)}\gamma_i f''(\gamma_i)]/f'(\gamma_i),$$

$$e_1 = g(0)/f'(0), \quad e_2 = e_1[g'(0) - \tfrac{1}{2} e_1 f''(0)]/f'(0),$$

where

$$f(x) = x(x-\gamma_1)(x-\gamma_2)(x-\gamma_3), \quad g(x) = (x-\kappa_1)(x-\kappa_2)(x-\kappa_3),$$

and $f' = df/dx$. If we put

$$s = v^{-1} + i\omega^{-1}q,$$

then v is the speed of propagation, and q is the attenuation coefficient of the wave. It follows from (5.6.50) and (5.6.55) that the low-frequency approximation to the propagation speeds and attenuation coefficients of the quasielastic and quasithermal modes are

$$v_i = V_i[1 - \tfrac{1}{8}(4c_2^{(i)} - 3c_1^{(i)2})\chi^2 + O(\chi^4)],$$

$$q_i = \frac{\omega^o}{2V_i} c_1^{(i)}(\chi^2 + O(\chi^4)),$$

$$v_4 = V_4(2e_1\chi)^{1/2}[1 + \frac{e_2}{2e_1}\chi + O(\chi^2)],$$

$$q_4 = \frac{\omega^o}{V_4}\left(\frac{1}{2e_1}\chi\right)^{1/2}\left[1 + \frac{e_2}{2e_1}\chi + O(\chi^2)\right],$$

where

$$V_i = (d\gamma_i/\rho)^{1/2}, \quad V_4 = (d/\rho)^{1/2}.$$

The quasielastic modes are wavelike and the quasithermal mode diffusive in type.

In [60], Chadwick has also studied the high-frequency behaviour and various generalized modes.

5.7 Isothermal acceleration waves

In Section 3.7 an acceleration wave was defined as a singular surface across which some partial derivatives of the motion $\underline{x} = \underline{x}(\underline{X},t)$ suffer jump discontinuities. Here we extend the definition by requiring that the thermal fields θ and η be continuous, while their first partial derivatives may exhibit jump discontinuities across the singular surface. There is a vast literature on the theory of acceleration waves in heat-conducting materials (see, for example, Coleman and Gurtin [84]. Truesdell [361], Chen [74], McCarthy [257]). Here we follow the work of Chadwick and Currie [56].

In a thermoelastic material it follows from the constitutive equations that the stress, entropy and internal energy are continuous and the heat flux vector is discontinuous on an acceleration wave S(t). It is known (see, for example, McCarthy [257]) that the energy equation implies

$$[Q_K]N_K = 0, \tag{5.7.1}$$

on S(t). Chadwick and Currie [56] have considered the equation

$$G_A\hat{Q}_A(E_{MN},T,G_K) = 0, \tag{5.7.2}$$

where $G_A = T_{,A}$, as a condition on the temperature gradient $\underline{G}$ in a given deformation-temperature state (E_{MN},T). When the heat-flux functions $\hat{Q}_A$ vanish identically in the state (E_{MN},T), the material is said to be a non-conductor. For a non-conductor, the equation (5.7.2) holds for all choices of $\underline{G}$. A material for which (5.7.2) has no non-zero solutions is said to be a normal conductor. An anomalous conductor is a material which is neither a non-conductor nor a normal conductor. This classification of materials into normal conductors, anomalous conductors and non-conductors is due to Chadwick and Currie [56]. In this section we examine some properties of acceleration waves travelling through homogeneous normal conductors. A detailed treatment of acceleration waves in the three types of materials previously defined is given in [56].

A region in which the material is homogeneously deformed, at rest, and at constant temperature is said to be a region of uniform state (cf. Chadwick and Currie [56]). We confine our attention to waves entering a region of uniform state.

Let

$$\underline{a} = [\ddot{\underline{x}}], \quad g = [T_{,K}]N_K, \quad h = [\eta_{,K}]N_K. \tag{5.7.3}$$

Here g and h are respectively thermal and entropic amplitudes of the wave. It follows from the compatibility conditions (3.7.14) that

$$\begin{aligned} &[x_{i,KL}] = U_N^{-2} a_i N_K N_L, \quad [\dot{x}_{i,K}] = -U_N^{-1} a_i N_K, \\ &[T_{,K}] = [G_K] = gN_K, \ [\dot{T}] = -U_N g, \ [\dot{\eta}] = -U_N h, \ [\eta_{,K}] = hN_K. \end{aligned} \tag{5.7.4}$$

An acceleration wave of zero thermal amplitude is called an isothermal wave and a wave of zero entropic amplitude an isentropic wave. We have the following theorem due to Chadwick and Currie [56].

<u>Theorem 5.7.1</u>. An acceleration wave propagating into a region of uniform state of a homogeneous normal conductor is necessarily isothermal.

<u>Proof</u>. Since $\underline{G}$ vanishes ahead of S(t), it follows from (4.1.17) that the heat flux is also identically zero in this region. The value of $\underline{G}$ on the rearward of S(t) is $-[\underline{G}]$ and the relations (5.7.1) and (5.7.4) imply

$$N_K \hat{Q}_K(E_{AB}, T, -gN_L) = 0.$$

The above relation leads to

$$-[G_K]\hat{Q}_K(E_{AB},T,-[G_L]) = 0. \tag{5.7.5}$$

Since the considered material is a normal conductor, the equation (5.7.5) holds only for $[\underline{G}] = 0$. From (5.7.4) we conclude that the acceleration wave is an isothermal wave. □

We consider the equations of motion in the form (1.1.6). In view of (4.1.10) and (4.1.15) we obtain

$$T_{Ai} = \frac{\partial\psi}{\partial x_{i,A}}, \quad \rho_o\eta = -\frac{\partial\psi}{\partial T}. \tag{5.7.6}$$

Since the considered material is homogeneous, the equations of motion can be written in the form

$$B_{MiKj}x_{j,MK} - \beta_{Mi}T_{,M} + \rho_o f_i = \rho_o \ddot{x}_i, \tag{5.7.7}$$

where

$$B_{MiKj} = \frac{\partial^2\psi}{\partial x_{i,M}\partial x_{j,K}}, \quad \beta_{Ai} = -\frac{\partial^2\psi}{\partial x_{i,A}\partial T}. \tag{5.7.8}$$

We introduce the notation

$$D_{MiKjRs} = \partial B_{MiKj}/\partial x_{s,R}. \tag{5.7.9}$$

We note that B_{AiKj}, β_{Ai} and D_{MiKjRs} are determined by the deformation-temperature state.

As in Section 3.7 we assume that the body force is continuous. On calculating the jump in (5.7.7) across the wave front, making use of the conditions (5.7.4) and the fact that the wave is isothermal, we obtain

$$(Q^o_{ij}(\underline{N}) - \rho_o U_N^2\,\delta_{ij})a_j = 0, \tag{5.7.10}$$

where

$$Q^{o}_{ij}(\underset{-}{N}) = B_{MiKj}\, N_M N_K. \tag{5.7.11}$$

The tensor $Q^{o}_{ij}(\underset{-}{N})$ is called the isothermal acoustic tensor for the direction $\underset{-}{N}$. In view of the symmetry of $Q^{o}_{ij}(\underset{-}{N})$ we deduce from (5.7.10) the following result (Truesdell [355]): the amplitude vector of an isothermal acceleration wave travelling in the direction of the unit vector $\underset{-}{N}$ is a proper vector of the isothermal acoustic tensor $Q^{o}_{ij}(\underset{-}{N})$ and the speed of propagation U_N is such that $\rho_o U_N^2$ is the corresponding proper number. Conversely, any proper vector of $Q^{o}_{ij}(\underset{-}{N})$ is a possible amplitude of an isothermal acceleration wave provided that the corresponding proper number is non-negative.

For the remainder of this section we confine our attention to plane acceleration waves. In order to derive the growth equation governing the variation of the acceleration amplitude, we first differentiate the equations (5.7.7) with respect to time,

$$D_{MiKjRs}\dot{x}_{s,R}x_{j,MK} + B_{MiKj}\dot{x}_{j,MK} - \left(\frac{\partial\beta_{Mi}}{\partial x_{j,B}}\dot{x}_{j,B} + \right.$$
$$\left. + \frac{\partial\beta_{Mi}}{\partial T}\dot{T}\right)T_{,M} - \beta_{Mi}\dot{T}_{,M} + \rho_o\dot{f}_i = \rho_o\dddot{x}_i. \tag{5.7.12}$$

Since the temperature and its first partial derivatives are continuous on an isothermal acceleration wave, we obtain from (3.7.14) the following compatibility conditions

$$[T_{,AB}] = PN_A N_B, \quad [\dot{T}_{,A}] = -U_N P N_A, \tag{5.7.13}$$

where

$$P = [T_{,KL}]N_K N_L. \tag{5.7.14}$$

It follows from (3.7.34), (3.7.35), (5.7.12) and (5.7.13) that

$$2\rho_o\frac{\delta_D a_i}{\delta t} - U_N^{-3} d_{isj} a_s a_j - \beta_{Mi} P U_N N_M = (Q^{o}_{ij}(\underset{-}{N}) - \rho_o U_N^2 \delta_{ij})c_j \tag{5.7.15}$$

where

$$d_{isj} = D_{MiKjRs}N_M N_K N_R. \tag{5.7.16}$$

We now consider the energy equation (4.1.16), rewritten as

$$\rho_0 T\dot{\eta} = \frac{\partial Q_A}{\partial x_{i,M}} x_{i,MA} + \frac{\partial Q_A}{\partial T} T_{,A} + \frac{\partial Q_A}{\partial T_{,B}} T_{,AB} + \rho_0 s. \tag{5.5.17}$$

We assume that the heat supply s is continuous.

In view of the uniform state assumption each term of equation (5.7.17) is zero ahead of S(t). Evaluation on the rearward side of S(t) therefore gives

$$\begin{aligned} -\rho_o TU_N h = [x_{i,MA}] \frac{\partial \tilde{Q}_A}{\partial x_{i,M}} (x_{r,L}, T, -gN_K) + \\ + [T_{,A}] \frac{\partial \tilde{Q}_A}{\partial T} (x_{r,M}, T, -gN_K) + \\ + PN_A N_B \frac{\partial \tilde{Q}_A}{\partial T_{,B}} (x_{r,M}, T, -gN_K). \end{aligned} \tag{5.7.18}$$

It follows from (4.1.17) that

$$\frac{\partial \tilde{Q}_A}{\partial x_{i,M}} = 0, \quad \frac{\partial \tilde{Q}_A}{\partial T} = 0 \text{ if } T_{,K} = 0. \tag{5.7.19}$$

Moreover, we have (Chadwick and Set [54])

$$\frac{\partial \tilde{Q}_A}{\partial T_{,B}} (x_{r,M}, T, 0) Y_A Y_B \geq 0 \text{ for all } Y_M. \tag{5.7.20}$$

For an isothermal acceleration wave, g = 0. The result of setting g = 0 in equation (5.7.18) and using (5.7.19) is

$$\rho_0 TU_N h = -\tilde{k}P, \tag{5.7.21}$$

where

$$\tilde{k} = \frac{\partial \tilde{Q}_A}{\partial T_{,B}} (x_{r,M}, T, 0) N_A N_B. \tag{5.7.22}$$

From the constitutive equations (4.1.15) we obtain

$$\rho_0 \dot{\eta} = -\frac{\partial^2\psi}{\partial T \partial x_{i,K}} \dot{x}_{i,K} - \frac{\partial^2\psi}{\partial T^2}\dot{T}. \tag{5.7.23}$$

Forming the jump of each term on S(t) and utilising the compatibility conditions (5.7.4) we obtain, for an isothermal acceleration wave,

$$h = (\rho_0 U_N^2)^{-1} \beta_{Ki} N_K a_i. \tag{5.7.24}$$

It follows from (5.7.20) and (5.7.22) that $\tilde{k} \geqq 0$. We assume that $\tilde{k} \neq 0$. Then (5.7.21) and (5.7.24) implies that

$$P = -(T/\tilde{k}U_N)\beta_{Ki}N_K a_i. \tag{5.7.25}$$

When the equation (5.7.15) is multiplied by a_i and use is made of the propagation condition (5.7.10), the symmetry of $Q^o_{ij}(\underline{N})$ and (5.7.25), we arrive at the following equation

$$2\rho_0 a_i \frac{\delta_D a_i}{\delta t} - U_N^{-3} d_{isj} a_i a_s a_j + (T/\tilde{k})(\beta_{Ki}N_K a_i)^2 = 0. \tag{5.7.26}$$

It is convenient to write

$$a_i = am_i, \tag{5.7.27}$$

where $\underline{m}$ is a unit proper vector of the isothermal acoustical tensor $Q^o_{ij}(\underline{N})$ associated with the proper number $\rho_0 U_N^2$ and a is called the strength of the wave. We note that

$$\delta_D/\delta t = U_N \, d/dN, \tag{5.7.28}$$

where N is distance measured along the normal to S(t). In view of (5.7.27) and (5.7.28) the equation (5.7.26) reduces to

$$\frac{da}{dN} = \alpha a^2 - \beta a, \tag{5.7.29}$$

where

$$\alpha = (1/2\ \rho_0 U_N^4) d_{isj} m_i m_s m_j,$$

$$\beta = (T/2\ \rho_0 \tilde{k} U_N)(\beta_{Ki} N_K m_i)^2.$$

We note that $\beta > 0$. The solution of the equation (5.7.29) is

$$a = a_0 e^{-\beta N}\{1 - (\alpha a_0/\beta)(1 - e^{-\beta N})\}^{-1},$$

where $a = a_0$ when $N = 0$. If $\alpha = 0$, then $a = a_0 \exp(-\beta N)$. We now assume that $\alpha \neq 0$. Then we have

(i) if $a_0 \operatorname{sgn} \alpha > \beta/|\alpha|$, a/a_0 increases monotonically, becoming infinite at time $-(1/\beta U_N)\log\{1-(\beta/\alpha a_0)\}$;

(ii) if $a_0 \operatorname{sgn} \alpha < \beta/|\alpha|$, a/a_0 decreases monotonically, approaching zero exponentially as $t \to \infty$;

(iii) if $a_0 \operatorname{sgn} \alpha = \beta/|\alpha|$, $a = a_0$ for $t \geqq 0$.

Detailed analyses of the growth of acceleration waves in heat-conducting elastic materials have been given by Chadwick and Currie [56], [58].

In the theory of elastic materials there are close parallels between acceleration waves and plane harmonic disturbances of a uniform equilibrium configuration. The relationships between harmonic and acceleration plane waves in the context of thermoelasticity has been studied by Chadwick [60]. It is shown that it is no longer the case that for a result referring to one kind of waves there is an associated result for waves of the other kind.

6 Thermoelastostatics

6.1 The body force analogy

In this section we consider the equilibrium theory of thermoelastic bodies. We assume that the primary state is a configuration of equilibrium at constant temperature. In absence of time dependence, the fundamental system of field equations describing the behaviour of a thermoelastic body, referred to the primary state, consists of the equations of equilibrium

$$s_{ji,j} + \rho F_i = 0, \tag{6.1.1}$$

the equilibrium energy equation

$$\phi_{i,i} + \rho S = 0, \tag{6.1.2}$$

and the constitutive equations

$$\begin{aligned} s_{ji} &= d_{ijrs}u_{r,s} - \beta_{ij}\theta, \\ \phi_i &= k_{ij}\theta_{,j}, \end{aligned} \tag{6.1.3}$$

on B. We consider the boundary conditions

$$\begin{aligned} &\underline{u} = \tilde{\underline{u}} \text{ on } \bar{S}_1, \quad \underline{s} = \tilde{\underline{p}} \text{ on } S_2, \\ &\theta = \tilde{\theta} \text{ on } \bar{S}_3, \quad q = \tilde{\phi} \text{ on } S_4. \end{aligned} \tag{6.1.4}$$

We assume that: (i) d_{ijrs}, β_{ij} and k_{ij} are continuously differentiable on $\bar{B}$ and $d_{ijrs} = d_{rsij}$, $\beta_{ij} = \beta_{ji}$; (ii) ρ is continuous and strictly positive on $\bar{B}$; (iii) $\underline{F}$ and S are continuous on $\bar{B}$; (iv) $\tilde{\underline{u}}$ is continuous on $\bar{S}_1$, and $\tilde{\theta}$ is continuous on $\bar{S}_3$; (v) $\tilde{\underline{p}}$ is piecewise regular on S_2, and $\tilde{\phi}$ is piecewise regular on S_4.

Substitution of (6.1.3) into the equations (6.1.1), (6.1.2) and (6.1.4) yields the equations

$$(d_{ijrs}u_{r,s})_{,j} - (\beta_{ij}\theta)_{,j} + \rho F_i = 0, \tag{6.1.5}$$

$$(k_{ij}\theta_{,j})_{,i} = -\rho S, \tag{6.1.6}$$

on B, and the boundary conditions

$$u_i = \tilde{u}_i \text{ on } \bar{S}_1, \quad (d_{ijrs}u_{r,s} - \beta_{ij}\theta)n_j = \tilde{p}_i \text{ on } S_2, \tag{6.1.7}$$

$$\theta = \tilde{\theta} \text{ on } \bar{S}_3, \quad k_{ij}\,\theta_{,j}n_i = \tilde{\phi} \qquad \text{on } S_4. \tag{6.1.8}$$

We note that the above system is uncoupled in the sense that the temperature field can be found by solving the heat flow problem (6.1.6), (6.1.8). In this chapter, we shall treat the temperature field as having already been so determined. The ordered array $\Lambda = (F_i, \tilde{u}_i, \tilde{p}_i, \theta)$ is called external data system on $\bar{B}$ for the mixed problem of thermoelastostatics. By a solution of the mixed problem of thermoelastostatics we mean an admissible displacement field $\underline{u}$ on $\bar{B}$ (cf. Sect. 2.1), that satisfies the equations (6.1.5) and the boundary conditions (6.1.7).

It follows from (6.1.5) and (6.1.7) that $\underline{u}$ is a solution of the mixed problem of thermoelastostatics corresponding to the external data system $\Lambda = (F_i, \tilde{u}_i, \tilde{p}_i, \theta)$ on $\bar{B}$ if and only if $\underline{u}$ is a solution of the mixed problem of elastostatics corresponding to the external data system $L = (F_i - (1/\rho)(\beta_{ij}\theta)_{,j}, \tilde{u}_i, \tilde{p}_i + \beta_{ij}\theta n_j)$ on $\bar{B}$. Following the classical thermoelastostatics, we refer to this result as the body force analogy (see, for example, Carlson [47], Sect.9). Thus, for every theorem in elastostatics, there is an associated result in the equilibrium theory of linear thermoelasticity. The body force analogy can be used to obtain the thermoelastic versions of the theorems from Sections 2.2 and 2.3.

6.2 Thermoelastic stresses in homogeneously deformed bodies

We assume throughout this section that the body B_o is homogeneous and isotropic. We suppose that the primary state is obtained from B_o by the isothermal homogeneous deformation (5.6.1). In addition, we assume throughout this section that the body forces F_i and the heat supply S are zero. We first establish a solution of the displacement-temperature equations of equilibrium. Then, this solution is used to solve a problem concerning with a penny-shaped crack in an infinite medium. The results we give here are due to England and Green [111].

The special expressions assumed by the relations (6.1.3) have already been given in (5.6.2). In absence of body forces and heat supply, the equations (6.1.5) and (6.1.6) reduce to

$$a_{11}u_{1,11} + \frac{1}{2}(a_{11}-d_{12})u_{1,22} + a_{54}u_{1,33} + \frac{1}{2}(a_{11}+d_{12})u_{2,12} + $$
$$+ (d_{13}+a_{45})u_{3,13} - \beta_1\theta_{,1} = 0,$$

$$\frac{1}{2}(a_{11}+d_{12})u_{1,12} + \frac{1}{2}(a_{11}-d_{12})u_{2,11} + a_{11}u_{2,22} + a_{54}u_{2,33} +$$
$$+ (d_{13}+a_{45})u_{3,23} - \beta_1\theta_{,2} = 0, \qquad (6.2.1)$$

$$(d_{13}+a_{45})u_{\alpha,\alpha 3} + a_{45}\Delta_o u_3 + a_{33}u_{3,33} - \beta_3\theta_{,3} = 0,$$

and

$$k_1\Delta_o\theta + k_3\theta_{,33} = 0, \qquad (6.2.2)$$

respectively. The constants a_{rs} $(r,s = 1,2,\ldots,6)$, β_1,β_3, d_{12} and d_{13} are given by (2.4.12), (2.4.14) and (5.6.3).

Since θ is assumed to be known we first seek a particular solution of the equations (6.2.1) which expresses u_i in terms of θ. The general solution is then found by adding to this particular solution the general solution of (6.2.1) when $\theta = 0$.

We seek a particular solution in the form (England and Green [111])

$$u_\alpha = r_1G_{,\alpha}, \quad u_3 = r_3G_{,3}, \qquad (6.2.3)$$

where G is a class C^2 field on B. The equations (6.2.1) are satisfied if

$$r_1 a_{11}\Delta_0 G + [r_1 a_{54} + r_3(d_{13}+a_{45})]G_{,33} - \beta_1\theta = 0,$$

$$[r_1(d_{13}+a_{45}) + r_3 a_{45}]\Delta_0 G + r_3 a_{33} G_{,33} - \beta_3\theta = 0. \qquad (6.2.4)$$

Recalling (6.2.2) we assume that G is any particular solution of the equations

$$k_1\Delta_0 G + k_3 G_{,33} = 0, \quad G_{,33} = \theta. \qquad (6.2.5)$$

We suppose that $k_1 \neq 0$, $k_2 \neq 0$. Equations (6.2.4) and (6.2.5) are compatible if

$$(a_{54}-a_{11}b)r_1 + (d_{13}+a_{45})r_3 = \beta_1,$$

$$-(d_{13}+a_{45})br_1 + (a_{33}-a_{45}b)r_3 = \beta_3,$$

where $b = k_3/k_1$. We obtain

$$a_{11}a_{45}(m_1-b)(m_2-b)r_1 = \beta_1(a_{33}-a_{45}b) - \beta_3(d_{13}+a_{45}),$$

$$a_{11}a_{45}(m_1-b)(m_2-b)r_3 = \beta_3(a_{54}-a_{11}b) + \beta_1 b(d_{13}+a_{45}), \qquad (6.2.6)$$

where m_α are the roots of the equation (2.5.2). We assume that a_{11} and a_{45} are non-zero. The particular solution is degenerate when $m_1 = b$ or $m_2 = b$. In this case there is a particular solution of the form

$$u_\alpha = d_1 x_3 G_{,\alpha 3}, \quad u_3 = d_2 x_3 G_{,33} + d_3 G_{,3}, \qquad (6.2.7)$$

where G is any particular integral of (6.2.5) and d_i are given by

$$D_1 d_1 = k_1[(d_{13} + a_{45})k_1\beta_3 - (k_1 a_{33} - k_3 a_{45})\beta_1],$$

$$D_1 d_2 = k_1[(k_3 a_{11} - k_1 a_{54})\beta_3 - (d_{13} + a_{45})k_3\beta_1],$$

$$(d_{13}+a_{45})D_1 d_3 = \beta_1(a_{11}k_3-a_{54}k_1)(k_1 a_{33}+k_3 a_{45}) - \qquad (6.2.8)$$

$$- k_1\beta_3(d_{13}+a_{45})(k_1a_{54}+k_3a_{11}),$$

$$D_1 = 2(k_3^2a_{11}a_{45} - k_1^2a_{33}a_{54}).$$

When the equation (2.5.2) has equal roots, the particular solution (6.2.7) degenerates. In this case there is a particular solution of the form

$$u_\alpha = e_1x_3^2G_{,\alpha 3}, \quad u_3 = e_2x_3^2G_{,33} + e_3x_3G_{,3} + e_4G_{,3} \tag{6.2.9}$$

where G is any particular integral of (6.2.5) and

$$8a_{33}a_{54}de_1 = -(d_{13}+a_{45})(\beta_3d - \beta_1b), \quad d = a_{54}/a_{45},$$

$$e_2 = de_1, \quad 4a_{33}a_{54}dk_1e_3 = (k_1a_{54}+k_3a_{11})(\beta_3d-\beta_1b),$$

$$4(d_{13}+a_{45})a_{54}k_3e_4 = 4a_{54}k_3\beta_1 - (d_{13}+a_{45})dk_1(\beta_3d-\beta_1b).$$

Application. In what follows we use the solutions (2.5.4) and (6.2.3) to solve a problem concerning with a penny-shaped crack in an infinite medium.

We deal only with non-degenerate case, since the solutions in the other cases may be found from this by a limiting process.

The result we give here is due to England and Green [111]. We consider an infinite homogeneous and isotropic medium which is subjected to the isothermal deformation (5.6.1) in which the stress component t_{33} is zero. We assume that in B_0 there is a penny-shaped crack of radius a_0 in the plane $X_3 = 0$ which becomes a crack of radius a in the same plane, where $a = \lambda_1 a_0$. We suppose that the equal distributions of temperature are applied to both surfaces of the crack. This problem is equivalent to the problem for the half-space $x_3 \geqq 0$, which is obtained from the unstrained body by the finite deformation (5.6.1), and then subjected to the following conditions for $x_3 = 0$,

$$s_{33} = 0, \quad 0 \leqq r \leqq a, \quad u_3 = 0, \quad a < r, \quad s_{3\alpha} = 0, \quad r \geqq 0, \tag{6.2.10}$$

$$\theta = f(r), \quad 0 \leqq r \leqq a, \qquad \theta_{,3} = 0, \quad r > a, \tag{6.2.11}$$

where f is a prescribed function, and $r = (x_\alpha x_\alpha)^{1/2}$. We impose the further condition that s_{ij} and θ vanish at infinity.

The solution of the boundary-value problem (6.2.2), (6.2.11) is (cf. Green and Zerna [140], p. 175)

$$\theta = \frac{1}{2}\int_{-a}^{a} [r^2 + (z + it)^2]^{-1/2} h(t)dt, \tag{6.2.12}$$

where $z = x_3(k_1/k_3)^{1/2}$, and h is an even function given by

$$h = \frac{2}{\pi}\frac{d}{dt}\int_0^t (t^2 - r^2)^{-1/2} rf(r)dr. \tag{6.2.13}$$

It follows from (6.2.5) and (6.2.12) that

$$G = \frac{k_3}{2k_1}\int_{-a}^{a} [(z+it)\ln\{z+it + [r^2 + (z + it)^2]^{1/2}\} - \\ - \{r^2 + (z + it)\}^{1/2}]h(t)dt. \tag{6.2.14}$$

Clearly, the second-order derivatives of G vanish at infinity. We seek the displacement field in the form

$$u_\alpha = (g_1+g_2)_{,\alpha} + r_1 G_{,\alpha}, \quad u_3 = (h_1 g_1 + h_2 g_2)_{,3} + r_3 G_{,3}, \tag{6.2.15}$$

where g_α and G satisfy the equations (2.5.5) and (6.2.5), respectively, and h_α, r_1, r_3 are given by (2.5.1) and (6.2.6). It follows from (5.6.2), (5.6.3) and (6.2.15) that

$$s_{33} = (c_{33}h_1 - c_{13}m_1)g_{1,33} + (c_{33}h_2 - c_{13}m_2)g_{2,33} + \\ + (c_{33}r_3 + c_{13}r_1 b)G_{,33} - \beta_3\theta, \tag{6.2.16}$$

$$s_{3\alpha} = c_{44}[(1+h_1)g_{1,3\alpha} + (1+h_2)g_{2,3\alpha} + (r_1+r_3)G_{,3\alpha}].$$

The boundary conditions (6.2.10) are satisfied if

$$(1+h_1)g_{1,3} + (1+h_2)g_{2,3} + (r_1+r_3)G_{,3} = 0, \; r \geqq 0,$$

$$h_1g_{1,3} + h_2g_{2,3} + r_3G_{,3} = 0, \quad r > a, \tag{6.2.17}$$

$$(c_{33}h_1-c_{13}m_1)g_{1,33} + (c_{33}h_2-c_{13}m_2)g_{2,33} +$$

$$+ (c_{33}r_3+c_{13}r_1b)G_{,33} = \beta_3\theta, \quad 0 \leqq r \leqq a,$$

for $x_3 = 0$. In view of equations (2.5.5) and (6.2.5) we put

$$g_1 = d_1G(x_1,x_2,v_1x_3) + C_1F(x_1,x_2,t_1x_3),$$

$$g_2 = d_2G(x_1,x_2,v_2x_3) - C_2F(x_1,x_2,t_2x_3), \tag{6.2.18}$$

where d_α are constants and

$$t_\alpha = m_\alpha^{-1/2}, \; v_\alpha = t_\alpha(k_3/k_1)^{1/2}, \quad C_\alpha^{-1} = t_\alpha(1 + h_\alpha). \tag{6.2.19}$$

The function F satisfies the equation

$$F_{,ii} = 0. \tag{6.2.20}$$

We choose the constants d_α so that

$$(1 + h_1)v_1d_1 + (1 + h_2)v_2d_2 = - (r_1+r_3),$$

$$h_1v_1d_1 + h_2v_2d_2 = - r_3. \tag{6.2.21}$$

Thus

$$d_1 = \frac{h_2r_1 - r_3}{v_1(h_1 - h_2)}, \qquad d_2 = \frac{r_3 - h_1r}{v_2(h_1 - h_2)} .$$

It follows from (6.2.18) and (6.2.21) that the boundary conditions (6.2.17) reduce to

$$F_{,3} = 0, \; r > a, \;\; F_{,33} = Af, \quad 0 \leqq r \leqq a, \tag{6.2.22}$$

on $x_3 = 0$, where

$$A = (\beta_3 - c_{33}r_3 - c_{13}r_1 b - A_1 d_1 v_1^2 - A_2 d_2 v_2^2)/(A_1 C_1 t_1^2 - A_2 C_2 t_2^2),$$

$$A_\alpha = c_{33}h_\alpha - c_{13}m_\alpha. \qquad (6.2.23)$$

The solution of the boundary-value problem (6.2.20), (6.2.22) is (cf. Green and Zerna [140], p. 174)

$$F = \frac{1}{2i}\int_{-a}^{a} g(t)\ln\{x_3 + it + [r^2 + (x_3 + it)^2]^{1/2}\}dt, \qquad (6.2.24)$$

where g is an odd function defined by

$$g = 2A\pi^{-1}\int_0^t sf(s)(t^2 - s^2)^{-1/2}\, ds,\ t \geqq 0. \qquad (6.2.25)$$

The corresponding stresses may be found by using the relations (5.6.2). We note that all stress components vanish at infinity. The normal displacement at the surface of the crack is given by

$$u_3 = h_1 g_{1,3} + h_2 g_{2,3} + r_3 G_{,3},\quad 0 \leqq r \leqq a,\ x_3 = 0. \qquad (6.2.26)$$

It follows from (6.2.18), (6.2.21) and (6.2.26) that

$$u_3 = (h_1 t_1 C_1 - h_2 t_2 C_2)F_{,3},\quad 0 \leqq r \leqq a,\ x_3 = 0. \qquad (6.2.27)$$

In the case when a constant temperature field is applied to the surfaces of the crack, so that $f = f_o$, where f_o is a constant, the relation (6.2.25) yields

$$g = 2Af_o\pi^{-1}t. \qquad (6.2.28)$$

It follows from (6.2.24), (6.2.27) and (6.2.28) that the normal displacement at the crack is given by

$$u_3 = -\ 2Af_o\pi^{-1}(h_1 t_1 C_1 - h_2 t_2 C_2)(a^2 - r^2)^{1/2},\quad 0 \leqq r \leqq a.$$

The solution of the problem may now be completed by straightforward calculation but details are left to the reader who may also refer to England and Green [111]. In the classical thermoelasticity, the problem of thermal stresses in an infinite elastic solid containing a penny-shaped crack was studied by Olesiak and Sneddon [280].

6.3 Thermal stresses in prestrained cylinders

In Section 2.7 we have studied the problem of torsion of a cylinder which has been subjected to a large extension from a natural state. In this section we study the deformation of the same cylinder under the action of a prescribed temperature distribution. We assume that the primary deformation is the isothermal deformation described in Section 2.7.

Since $t_{11} = t_{22} = 0$, the equations (5.6.2) reduce to

$$
\begin{aligned}
s_{11} &= c_{11}u_{1,1} + c_{12}u_{2,2} + c_{13}u_{3,3} - \beta_1\theta,\\
s_{22} &= c_{12}u_{1,1} + c_{11}u_{2,2} + c_{13}u_{3,3} - \beta_1\theta,\\
s_{33} &= c_{13}(u_{1,1} + u_{2,2}) + (c_{33}+t_{33})u_{3,3} - \beta_3\theta,\\
s_{23} &= c_{44}(u_{2,3}+u_{3,2}), \quad s_{32} = (c_{44}+t_{33})u_{2,3}+c_{44}u_{3,2},\\
s_{31} &= (c_{44}+t_{33})u_{1,3}+c_{44}u_{3,1}, \quad s_{13} = c_{44}(u_{1,3}+u_{3,1}),\\
s_{12} &= s_{21} = \tfrac{1}{2}(c_{11}-c_{12})(u_{1,2}+u_{2,1}).
\end{aligned}
\tag{6.3.1}
$$

We assume that the body forces are zero. The equations of equilibrium (6.1.1) become

$$s_{ji,j} = 0 \text{ on } B. \tag{6.3.2}$$

In the secondary deformation the cylinder is assumed to be free of lateral loading, so that the conditions on the lateral boundary are

$$s_{ji}n_j = 0 \text{ on } \Pi. \tag{6.3.3}$$

We assume that the loading applied on the end located at $x_3 = 0$ is equivalent to zero. Thus, for $x_3 = 0$ we have the conditions

$$\int_\Sigma s_{3\alpha}\, da = 0, \tag{6.3.4}$$

$$\int_\Sigma s_{33} da = 0, \quad \int_\Sigma (x_\alpha\, s_{33} + u_\alpha t_{33}) da = 0, \tag{6.3.5}$$

$$\int_\Sigma e_{3\alpha\beta}\, x_\alpha\, s_{3\beta}\, da = 0. \tag{6.3.6}$$

Throughout this section we assume that the temperature field θ has the form

$$\theta = f(x_1,x_2), \quad (x_1,x_2) \in \Sigma, \tag{6.3.7}$$

where f is given.

The problem consists in the determination of an admissible displacement field $\underline{u}$ that satisfies the equations (6.3.1), (6.3.2) and the conditions (6.3.3)-(6.3.6) assuming that the temperature θ has the form (6.3.7).

We seek a solution of the problem in the form [189]

$$u_\alpha = - a_3\, \nu\, x_\alpha + v_\alpha(x_1,x_2), \quad u_3 = a_3 x_3, \tag{6.3.8}$$

where ν is defined by (2.9.23), a_3 is an unknown constant and v_α are unknown functions. We introduce the notations

$$\begin{aligned} \overset{o}{s}_{11} &= c_{11}v_{1,1} + c_{12}v_{2,2} - \beta_1\theta, \\ \overset{o}{s}_{22} &= c_{12}v_{1,1} + c_{11}v_{2,2} - \beta_1\theta, \\ \overset{o}{s}_{12} &= \overset{o}{s}_{21} = \frac{1}{2}(c_{11} - c_{12})(v_{1,2} + v_{2,1}). \end{aligned} \tag{6.3.9}$$

It follows from (2.9.23), (6.3.1), (6.3.8) and (6.3.9) that

$$\begin{aligned} s_{\alpha\beta} &= \overset{o}{s}_{\alpha\beta}\,, \quad s_{\alpha 3} = s_{3\alpha} = 0, \\ s_{33} &= (c_{33}+t_{33}-2\nu c_{13})a_3 + c_{13}v_{\alpha,\alpha} - \beta_3 f. \end{aligned} \tag{6.3.10}$$

The equations of equilibrium (6.3.2) become

$$s^{o}_{\beta\alpha,\beta} = 0 \text{ on } \Sigma. \tag{6.3.11}$$

By (6.3.10), the boundary conditions (6.3.3) reduce to

$$s^{o}_{\beta\alpha}\, n_{\beta} = 0 \text{ on } L. \tag{6.3.12}$$

Thus, the functions v_{α} are the components of the displacement vector in the thermoelastic plane strain problem (6.3.9), (6.3.11), (6.3.12) corresponding to the temperature distribution f. In what follows we assume that this problem is solved. Clearly, by Theorem 2.3.1

$$v_{\alpha} = v^{*}_{\alpha} + a_{\alpha}, \tag{6.3.13}$$

where a_{α} are arbitrary constants, and v^{*}_{α} are given.

The conditions (6.3.4) and (6.3.6) are satisfied on the basis of the relations (6.3.10). In view of (6.3.10), the conditions (6.3.5) reduce to

$$\begin{aligned}
&(c_{33}+t_{33}-2\nu c_{13})Aa_{3} = \int_{\Sigma} (\beta_{3}f-c_{13}v^{*}_{\alpha,\alpha})da,\\
&t_{33}Aa_{\alpha} + [c_{33}-2\nu c_{13} + (1-\nu)t_{33}]x^{o}_{\alpha}Aa_{3} = \\
&\qquad = \int_{\Sigma} [x_{\alpha}(\beta_{3}f-c_{13}v^{*}_{\alpha,\alpha}) - t_{33}v^{*}_{\alpha}]da.
\end{aligned} \tag{6.3.14}$$

We assume that (2.9.26) holds. We conclude that the functions v_{α} are given by the two-dimensional boundary value problem (6.3.9), (6.3.11), (6.3.12) and the constants a_{i} are characterized by (6.3.14).

We now assume that $f = T^{o} = \text{const}$. Then, the boundary value problem (6.3.9), (6.3.11), (6.3.12) implies that

$$v^{*}_{\alpha} = Cx_{\alpha},$$

where

$$C = \beta_{1}T^{o}/(c_{11} + c_{12}).$$

Assume that $x_\alpha^o = 0$. Then (6.3.14) implies

$$a_\alpha = 0, \quad a_3 = (\beta_3 - 2Cc_{13})T^o/[A(c_{33}+t_{33}-2\nu c_{13})].$$

6.4 Non-uniformly heated bodies

α) Torsion of an initially heated cylinder. In this section we apply the theory of Section 4.5 to study some boundary-value problems for bodies with infinitesimal initial deformations. We suppose that the region B_0 from here on refers to the interior of a right solid or hollow circular cylinder of length ℓ with the open cross-section Σ_o and the lateral boundary Π_o. The rectangular Cartesian coordinate frame is supposed to be chosen in such a way that the X_3-axis coincides with the line of centroids of cross sections and X_1OX_2-plane contains one of the terminal cross-sections, while the other is in the plane $X_3 = \ell$. We denote by Σ_1^o and Σ_2^o, respectively, the cross-section located at $X_3 = 0$ and $X_3 = \ell$. Let Σ_α be the image of the surface Σ_α^o in the configuration B.

We assume that the body B_0 is homogeneous and isotropic. The cylinder is supposed to be free from lateral loading. Moreover, we assume that the body forces are absent.

Throughout this section, we shall refer the secondary deformation to the configuration B. It is well-known that in the linear theory is no difference between Eulerian and Lagrangian descriptions. Since we are considering the primary state to be a small deformation from the reference state B_0, in the functions v_i, ζ, e_{ij}, and t_{ij} the coordinates X_A may be replaced by x_s to the degree of approximation involved. Thus, the first deformation is governed by the equilibrium equations

$$t_{ji,j} = 0, \tag{6.4.1}$$

the energy equation

$$q_{i,i} = -\rho_0 s, \tag{6.4.2}$$

and the constitutive equations

$$t_{ij} = \lambda e_{rr}\delta_{ij} + 2\mu e_{ij} - \beta\zeta\delta_{ij}, \quad q_i = k\zeta_{,i}, \tag{6.4.3}$$

on B. We consider the following mechanical boundary conditions

$$t_{\alpha i}n_{\alpha} = 0 \text{ on } \Pi, \; t_{3\alpha} = 0, \quad v_3 = 0 \text{ on } \Sigma_{\alpha}, \tag{6.4.4}$$

where Π is the image of the surface Π_0 in the configuration B, and the thermal boundary condition

$$q_i n_i = 0 \text{ on } \Sigma_{\alpha}. \tag{6.4.5}$$

To consider a class of initial deformations we assume that the heat supply and the thermal variable on Π are prescribed in such a way that the temperature field ζ has the form

$$\zeta = \zeta(r), \tag{6.4.6}$$

where $r^2 = x_{\alpha} x_{\alpha}$. We now present two simple examples in which the temperature ζ has the form (6.4.6).

i) Let B_0 be a right hollow cylinder whose cross-section Σ_0 is bounded by two concentric circles of radius a_1 and a_2 ($a_1 < a_2$). We assume that the temperature field ζ on the inner surface of the cylinder is T_1 and that the temperature on the outer surface is T_2, where T_{α} are constants. Moreover, we assume that $s = 0$. In this case, from (6.4.2), (6.4.3) and (6.4.5) we obtain

$$\zeta = C_1 \ln r + C_2, \tag{6.4.7}$$

where

$$C_1 = \frac{T_2 - T_1}{\ln(a_2/a_1)}, \quad C_2 = \frac{T_1 \ln a_2 - T_2 \ln a_1}{\ln(a_2/a_1)}.$$

ii) Let Σ_0 be a simple connected region bounded by a circle of radius a. We assume that the temperature on the lateral surface is A, and that the heat supply is Q, where A and Q are given constants. Then,

$$\zeta = -\frac{1}{4k}\rho_0 Q(r^2 - a^2) + A.$$

It follows from (6.4.1), (6.4.4) and (6.4.6) that the body B is in a state of thermoelastic plane strain, parallel to the X_1OX_2-plane. The displacement field corresponding to the temperature field (6.4.6) has the following form (see, for example, Timoshenko and Goodier [347], Sect. 135)

$$v_\alpha = x_\alpha \psi(r), \quad v_3 = 0, \tag{6.4.8}$$

where ψ depends only on r. Clearly, if ζ depends only on r, then the conditions (6.4.5) are satisfied.

It follows from (6.4.3) and (6.4.8) that

$$\begin{aligned} &\varepsilon_{\alpha\beta} = v_{\alpha,\beta} = \psi\delta_{\alpha\beta} + \frac{1}{r} x_\alpha x_\beta \psi', \\ &\varepsilon_{i3} = 0,\ \varepsilon_{\rho\rho} = 2\psi + r\psi',\ t_{\alpha\beta} = \lambda\varepsilon_{\rho\rho}\delta_{\alpha\beta} + 2\mu\varepsilon_{\alpha\beta} - \beta\zeta\delta_{\alpha\beta}, \\ &\quad t_{\alpha 3} = 0,\ t_{33} = \lambda\varepsilon_{\rho\rho} - \beta\zeta, \end{aligned} \tag{6.4.9}$$

where $\psi' = d\psi/dr$.

If we consider the example (i) where the temperature ζ has the form (6.4.7), then the function ψ is given by

$$\psi = G_1 + \frac{1}{r^2} G_2 + \frac{\beta}{\lambda + 2\mu}\psi_0(r), \tag{6.4.10}$$

where

$$\psi_0(r) = r^{-2}[f(r)-f(a_1)], \quad f(r) = \frac{1}{2}[\zeta(r) - \frac{1}{2}c_1]r^2,$$

$$a_1^2(\lambda + \mu)G_1 = \mu G_2 = \frac{a_1^2\mu\beta}{(\lambda + 2\mu)(a_2^2-a_1^2)}[f(a_2)-f(a_1)].$$

Clearly, by (6.4.8) the cylinder B_0 is deformed into a right circular cylinder of length ℓ with the ends Σ_1 and Σ_2, located in the plane $X_3 = 0$ and $X_3 = \ell$, respectively. We suppose that the secondary state B is a configuration of equilibrium which is obtained from B by a torsion about the X_3-axis. We assume that in the secondary deformation the body is free of lateral loading and that $F_i = S = 0$. The equations (6.1.1) and (6.1.2) become

$$s_{ji,j} = 0, \quad \phi_{i,i} = 0 \text{ on } B. \tag{6.4.11}$$

We consider the temperature condition

$$\theta = 0 \quad \text{on} \quad \partial B. \tag{6.4.12}$$

On the lateral boundary we have the conditions

$$s_{\alpha i} n_\alpha = 0 \text{ on } \Pi. \tag{6.4.13}$$

In the primary deformation the resultant force and the resultant moment about 0 of the tractions acting on Σ_1 are denoted by $\underline{R}$ and $\underline{M}$, respectively. We have

$$R_i = -\int_{\Sigma_1} t_{3i} da, \quad M_\alpha = \int_{\Sigma_1} e_{3\beta\alpha} x_\beta t_{33} da, \quad M_3 = -\int_{\Sigma_1} e_{3\alpha\beta} x_\alpha t_{3\beta} da.$$

In the secondary deformation, the resultant force and the resultant moment about 0 of the tractions acting on the image of Σ_1 are denoted by $\underline{R}^*$ and $\underline{M}^*$, respectively. It follows that

$$R_i^* = -\int_{\Sigma_1} (t_{3i}+s_{3i}) da, \quad M_\alpha^* = e_{3\beta\alpha} \int_{\Sigma_1} [(x_\beta+u_\beta)(t_{33}+s_{33}) - u_3(t_{3\beta}+s_{3\beta})] da,$$

$$M_3^* = -\int_{\Sigma_1} e_{3\alpha\beta}(x_\alpha+u_\alpha)(t_{3\beta} + s_{3\beta}) da.$$

Since B^* is obtained from B by a torsion about X_3-axis, we have the conditions

$$\underline{R}^* = \underline{R}, \quad \underline{M}^* - \underline{M} = (0,0,M_3),$$

where M_3 is a given constant. These conditions can be written in the form

$$\int_{\Sigma_1} s_{3\alpha} \, da = 0, \tag{6.4.14}$$

$$\int_{\Sigma_1} s_{33}da = 0, \tag{6.4.15}$$

$$\int_{\Sigma_1} (x_\beta s_{33} + u_\beta t_{33} - u_3 t_{3\beta})da = 0, \tag{6.4.16}$$

$$\int_{\Sigma_1} e_{3\alpha\beta}(x_\alpha s_{3\beta} + u_\alpha t_{3\beta})da = -M_3, \tag{6.4.17}$$

to the degree of approximation involved.

The problem consists in finding of the functions u_i, $\theta \in C^2(B) \cap C^1(\bar{B})$, satisfying the equations (4.2.20), (4.2.32), (6.4.11) and the conditions (6.4.12)-(6.4.17), where C_{ijrs}, β_{ij}, h_{ijr}, a_i and k_{ij} are given by (4.5.12).

We seek the solution of the problem in the form

$$u_\alpha = \tau e_{3\beta\alpha}x_\beta x_3, \; u_3 = \tau\phi(x_1,x_2), \; \theta = 0 \text{ on } B, \tag{6.4.18}$$

where τ is an unknown constant, and ϕ is an unknown function. Then, we have

$$u_{\alpha,\beta} = \tau e_{3\beta\alpha}x_3, \; u_{\alpha,3} = \tau e_{3\beta\alpha}x_\beta, \; u_{3,3} = 0, \; u_{3,\alpha} = \tau\phi_{,\alpha}. \tag{6.4.19}$$

In view of (6.4.18) and (6.4.19), the constitutive coefficients which are of interest are $C_{ij\alpha3}$ and $h_{i\beta3}$. By (4.5.12) and (6.4.9) we have

$$C_{\alpha\beta\rho3} = 0, \; C_{33\alpha3} = 0, \; C_{\alpha3\beta3} = A_{\alpha\beta}, \; h_{i\alpha3} = k_2\delta_{i3}\zeta_{,\alpha}, \tag{6.4.20}$$

where

$$A_{\alpha\beta} = A_{\beta\alpha} = [\mu+(\nu_2-\mu)\varepsilon_{\rho\rho}-\beta_2\zeta]\delta_{\alpha\beta} + 2(\mu+\nu_3)v_{\alpha,\beta}. \tag{6.4.21}$$

It follows from (4.2.20), (4.2.32), (4.5.12), (6.4.9), (6.4.18) and (6.4.20) that

$$s_{11} = t_{12}u_{1,2}, \; s_{12} = t_{11}u_{2,1}, \; s_{21} = t_{22}u_{1,2},$$

$$s_{22} = t_{21}u_{2,1}, \; s_{33} = 0,$$

$$s_{\alpha 3} = \tau[A_{\alpha\rho} + t_{\alpha\rho})\phi_{,\rho} - A_{\alpha\rho}e_{3\rho\nu}x_{\nu}], \tag{6.4.22}$$

$$s_{3\alpha} = \tau(A_{\alpha\rho}\phi_{,\rho} - A_{\alpha\rho}e_{3\rho\nu}x_{\nu} - e_{3\alpha\rho}x_{\rho}t_{33}),$$

$$\phi_{\alpha} = 0, \ \phi_3 = \tau k_2\zeta_{,\alpha}\phi_{,\alpha}.$$

Since ϕ_3 is independent of x_3, the last of the equations (6.4.11) is satisfied. In view of (6.4.19) and (6.4.22), we have

$$s_{ji,j} = s_{\rho 1,\rho} = u_{1,2}t_{\alpha 2,\alpha}, \ s_{j2,j} = u_{2,1}t_{\alpha 1,\alpha}. \tag{6.4.23}$$

Since the initial stresses t_{ij} satisfy the equations $t_{\alpha\beta,\alpha} = 0$, we conclude from (6.4.23) that the first two of the equilibrium equations (6.4.11) are satisfied. Further, since

$$A_{\alpha\rho,\alpha} = \frac{1}{r} x_{\rho} \frac{d}{dr} [\mu + (\nu_2 + \nu_3)\varepsilon_{\alpha\alpha} - \beta_2\zeta],$$

$$A_{\alpha\rho,\alpha}e_{3\rho\nu}x_{\nu} = 0, \qquad A_{\alpha\rho}e_{3\rho\alpha} = 0,$$

it follows that the third of the equilibrium equations (6.4.11) reduces to

$$((A_{\alpha\rho} + t_{\alpha\rho})\phi_{,\rho})_{,\alpha} = 0 \text{ on } \Sigma_1. \tag{6.4.24}$$

By (6.4.4) and (6.4.22) we obtain

$$s_{\alpha 1}n_{\alpha} = u_{1,2}t_{\rho 2}n_{\rho} = 0, \ \ s_{\alpha 2}n_{\alpha} = u_{2,1}t_{\beta 1}n_{\beta} = 0,$$

so that the first two of the boundary conditions (6.4.13) are satisfied. With the aid of (6.4.9) and (6.4.21) we find that

$$A_{\alpha\rho} = P\delta_{\alpha\rho} + \frac{2}{r}(\mu + \nu_3)x_{\alpha}\, x_{\rho}\psi', \tag{6.4.25}$$

where

$$P = \mu + (\nu_2 - \mu)\varepsilon_{\alpha\alpha} - \beta_2\zeta + 2(\mu + \nu_3)\psi. \tag{6.4.26}$$

Since

$$e_{3\rho\nu}x_\nu n_\rho = 0 \text{ on } \Pi,$$

from (6.4.25) we obtain

$$A_{\alpha\rho}e_{3\rho\nu}x_\nu n_\alpha = 0 \text{ on } \Pi.$$

Thus, the third of the boundary conditions (6.4.13) reduces to

$$(A_{\alpha\beta} + t_{\alpha\beta})\phi_{,\beta}n_\alpha = 0 \quad \text{on } \partial\Sigma_1. \tag{6.4.27}$$

It follows from (6.4.24) and (6.4.27) that $\phi = 0$ is a solution of the corresponding boundary value problem.

The conditions (6.4.15) and (6.4.16) are identically satisfied in view of (6.4.9), (6.4.18) and (6.4.22).

From (6.4.22) and (6.4.25) we have

$$s_{\alpha 3} = -\tau P e_{3\alpha\nu}x_\nu, \quad s_{3\alpha} = -\tau e_{3\alpha\nu}x_\nu(P+t_{33}), \quad \phi_3 = 0. \tag{6.4.28}$$

Since $P + T_{33}$ depends only on r,

$$\int_{\Sigma_1} x_\nu(P + t_{33})da = 0,$$

so that the conditons (6.4.14) are satisfied.

It follows from (6.4.9), (6.4.17) and (6.4.28) that

$$\tau D = -M_3, \tag{6.4.29}$$

where

$$D = \int_{\Sigma_1} (P + t_{33})r^2 da, \tag{6.4.30}$$

is the torsional rigidity of the heated cylinder. If $D \neq 0$, then the constant τ is determined by (6.4.29). We note that the equations

$$s_{ji,j} = 0, \quad s_{ji} = d_{ijrs}u_{r,s},$$

and the divergence theorem imply that

$$\int_B d_{ijrs}u_{i,j}u_{r,s}dv = \int_{\partial B} s_{ji}n_ju_ida. \tag{6.4.31}$$

We assume that d_{ijrs} are positive definite in the sense that there exists a positive constant d_o such that

$$\int_B d_{ijrs}\xi_{ij}\xi_{rs}dv \geqq d_o \int_B \xi_{ij}\xi_{ij}dv, \tag{6.4.32}$$

for all tensors ξ_{ij}. Then, we can prove that $D > 0$. Indeed, by (6.4.13) and (6.4.18),

$$\int_{\partial B} s_{ji}n_ju_ida = \int_{\Sigma_2} s_{3i}u_ida = \tau\ell\int_{\Sigma_2} e_{3\alpha\beta}x_\alpha s_{3\beta}da = \tau^2\ell D. \tag{6.4.33}$$

It follows from (6.4.31)-(6.4.33) that $D > 0$.

We conclude that the solution of the torsion problem is given by

$$u_\alpha = \tau e_{3\beta\alpha}x_\beta, \quad u_3 = 0, \; \theta = 0,$$

where τ is defined by (6.4.29).

We now consider two special cases.

a) Assume that $\zeta = T_1$, where T_1 is a given constant. Then,

$$v_\alpha = Cx_\alpha, \quad v_3 = 0,$$

where

$$C = \frac{\beta T_1}{2(\lambda + \mu)} .$$

In this case,

$$P + t_{33} = \mu + \kappa T_1,$$

with

$$\kappa = \frac{1}{\lambda + \mu}(\nu_2 + \nu_3 + \lambda)\beta - (\beta + \beta_2).$$

From (6.4.30), we obtain

$$D = (\mu + \kappa T_1)I,$$

where I is the polar moment of inertia of the cross-section Σ_1,

$$I = \int_{\Sigma_1} r^2 da.$$

If B_0 is a solid circular cylinder of radius a, then the cylinder B has the radius b, given by

$$b = a|1 + C|.$$

Clearly, $I = \pi b^4/2$.

b) In the case of example (i) where the temperature ζ has the form (6.4.7), we have

$$P + t_{33} = \mu + (\nu_2 + \lambda - \mu)r\psi' + 2(\lambda + \nu_2 + \nu_3)\psi - (\beta + \beta_2)\zeta.$$

It follows from (6.4.8), (6.4.10) and (6.4.30) that

$$D = 2\pi[F(b_2) - F(b_1)],$$

where

$$F(r) = \frac{1}{4} C_1\kappa_1 r^4 \ln r - \kappa_2 r^6 + \frac{1}{4}\kappa_3 r^4 + \kappa_4 r^2,$$

$$\kappa_1 = \beta^*(\lambda + \nu_2 + \nu_3) - \beta - \beta_2, \; \beta^* = \beta/(\lambda + 2\mu),$$

$$\kappa_2 = \frac{1}{3}\beta^*(\lambda + \nu_2 - \mu)f(a_1),$$

$$\kappa_3 = \mu + (\lambda + \nu_2 + \nu_3)(2G_1 - \beta^* C_1 + \beta^* C_2) + \frac{1}{2} C_1(\lambda + \nu_2 - \mu)\beta^* -$$

$$- (\beta + \beta_2)(C_2 - \frac{1}{4} C_1),$$

$$\kappa_4 = G_2(\mu + \nu_3) - \beta^*(\lambda + \nu_2 + \nu_3)f(a_1), \quad b_\alpha = a_\alpha|1 + \psi(a_\alpha)|.$$

It is a simple matter to verify that if $\zeta = 0$, then $\kappa_1 = \kappa_2 = \kappa_4 = 0$, $\kappa_3 = \mu$, $b_\alpha = a_\alpha$, and D reduces to the classical torsional rigidity for the unheated cylinder.

β) Thermoelastic bodies with constant initial heat flux vector.

We now assume that B_o is homogeneous and isotropic, and refer the secondary deformation to the configuration B_o. We use the theory of the Section 4.5 to obtain the basic equations when the state B is obtained from the state B_o by a small deformation characterized by the temperature difference field

$$\zeta = T_1 + T_2 X_3, \tag{6.4.34}$$

where T_α are constants. The corresponding displacement field is given by (see, for example, Carlson [47], Sect. 15)

$$V_A = \alpha[(T_1 + T_2 X_3)X_A - \frac{1}{2} T_2 X_R X_R \delta_{3A}] + v_A^o, \tag{6.4.35}$$

where

$$\alpha = \frac{\beta}{3\lambda + 2\mu},$$

and v_A^o is an arbitrary rigid displacement field.

It follows from (4.5.5), (4.5.11), (6.4.34) and (6.4.35) that

$$\varepsilon_{AB} = \alpha(T_1 + T_2 X_3)\delta_{AB}, \quad T_{AB} = 0. \tag{6.4.36}$$

Then, by (4.5.11), (6.4.35) and (6.4.36) we find

$$D_{iAjN} = (\lambda + 3\alpha\nu_1\zeta - \beta_1\zeta)\delta_{iA}\delta_{jN} + (\mu + 3\nu_2\alpha\zeta - \beta_2\zeta)(\delta_{jA}\delta_{iN} +$$

$$+ \delta_{ij}\delta_{AN}) + (\lambda v_{M,N} + 2\nu_2\alpha\zeta\delta_{MN})\delta_{iA}\delta_{jM} + (\lambda v_{L,A} +$$

$$+ 2\alpha\nu_2\zeta\delta_{AL})\delta_{iL}\delta_{jN} + (\mu v_{L,N} + 2\alpha\nu_3\zeta\delta_{LN})\delta_{jA}\delta_{iL} +$$

$$+ (\mu v_{M,A} + 2\nu_3\alpha\zeta\delta_{AM})\delta_{iN}\delta_{jM} + 2(\mu + \nu_3)\alpha\zeta\delta_{AN}\delta_{ij} +$$

$$+ 2\nu_3\alpha\zeta\delta_{ij}\delta_{AN}, \tag{6.4.37}$$

$$F_{Ai} = [\beta + (3\beta_1 + 2\beta_2)\alpha\zeta + d\zeta]\delta_{iA} + \beta\delta_{iM}v_{M,A}, \; A = a_o + 3\alpha\zeta d + h\zeta,$$

$$G_{AiK} = T_2[k_1\delta_{iK}\delta_{3A} + k_2(\delta_{iA}\delta_{3K} + \delta_{AK}\delta_{i3})], \quad D_A = gT_2\,\delta_{3A},$$

$$K_{MN} = [k + \alpha(k_1 + 4k_2)\zeta + g\zeta]\delta_{MN}.$$

The equations (4.2.17) and (4.2.28) become

$$S_{11} = (\lambda + \lambda_1\zeta)u_{R,R} + 2(\mu + \mu_1\zeta)u_{1,1} + \alpha T_2[(\lambda+\mu)X_1(u_{1,3} - u_{3,1}) +$$

$$+ \lambda X_2(u_{2,3} - u_{3,2})] - (\beta + \delta_1\zeta)\theta,$$

$$S_{22} = (\lambda + \lambda_1\zeta)u_{R,R} + 2(\mu + \zeta\mu_1)u_{2,2} +$$

$$+\alpha T_2[\lambda X_1(u_{1,3} - u_{3,1}) + (\lambda+\mu)X_2(u_{2,3} - u_{3,2})] - (\beta + \delta_1\zeta)\theta,$$

$$S_{33} = (\lambda + \zeta\lambda_1)u_{R,R} + 2\,(\mu + \mu_1\zeta)u_{3,3} +$$

$$+ (\lambda+\mu)\alpha T_2[X_1(u_{1,3} - u_{3,1}) + X_2(u_{2,3} - u_{3,2})] - (\beta + \delta_1\zeta)\theta,$$

$$S_{12} = (\mu + \mu_1\zeta)(u_{1,2} + u_{2,1}) + \mu\alpha T_2(X_2u_{1,3} - X_1u_{3,2}),$$

$$S_{21} = (\mu + \mu_1\zeta)(u_{1,2} + u_{2,1}) + \mu\alpha T_2(X_1u_{2,3} - X_2u_{3,1}),$$

$$S_{31} = (\mu + \mu_1\zeta)(u_{3,1} + u_{1,3}) + \alpha T_2X_1[(\lambda+\mu)u_{R,R} - \mu u_{2,2}] +$$
$$+ \mu\alpha T_2X_2u_{2,1} - \beta\alpha T_2X_1\theta, \tag{6.4.38}$$

$$S_{13} = (\mu+\zeta\mu_1)(u_{1,3} + u_{3,1}) - \alpha T_2 X_1[(\lambda+\mu)u_{R,R} - \mu u_{2,2}] -$$

$$- \mu\alpha T_2 X_2 u_{1,2} + \beta\alpha T_2 X_1 \theta,$$

$$S_{32} = (\mu+\mu_1\zeta)(u_{2,3} + u_{3,2}) + \alpha T_2 X_2[(\lambda+\mu)u_{R,R} - \mu u_{1,1}] +$$

$$+ \mu\alpha T_2 X_1 u_{1,2} - \beta\alpha T_2 X_2 \theta,$$

$$S_{23} = (\mu+\mu_1\zeta)(u_{2,3}+u_{3,2}) - \alpha T_2 X_2[(\lambda+\mu)u_{R,R} - \mu u_{1,1}] -$$

$$- \mu\alpha T_2 X_1 u_{2,1} + \beta\alpha T_2 X_2 \theta,$$

$$\eta = (\beta+\delta_1\zeta)u_{R,R} + (a_o+a_1\zeta)\theta + \alpha\beta T_2[X_1(u_{1,3}-u_{3,1}) +$$

$$+ X_2(u_{2,3}-u_{3,2})],$$

$$\Phi_\Lambda = T_2 k_2(u_{\Lambda,3} + u_{3,\Lambda}) + (k + \kappa\zeta)\theta_{,\Lambda},$$

$$\Phi_3 = T_2(k_1 u_{R,R} + 2k_2 u_{3,3}) + gT_2\theta + (k + \kappa\zeta)\theta_{,3},$$

where

$$\lambda_1 = 3\alpha\nu_1-\beta_1+2\alpha(\lambda+2\nu_2), \quad \mu_1 = 3\alpha\nu_2 - \beta_2 + 2\alpha(\mu + 2\nu_3),$$
$$\delta_1 = d + \alpha(\beta + 2\beta_2 + 3\beta_1), \quad \kappa = g + \alpha(k_1+4k_2), \quad a_1 = 3\alpha d + h. \tag{6.4.39}$$

Substitution of (6.4.38) into equations (1.2.34) and (4.2.38) yields the displacement-temperature equation of motion

$$L_\Lambda U + (\lambda+\mu)\alpha T_2[X_\Lambda u_{R,R3} + X_\Gamma(u_{\Gamma,3\Lambda}-u_{3,\Gamma\Lambda})] +$$

$$+ [(\lambda+\mu)\alpha+\mu_1]T_2 u_{\Lambda,3} - [(\lambda+2\mu)\alpha-\mu_1]T_2 u_{3,\Lambda} -$$

$$-\alpha\beta T_2 X_\Lambda \theta_{,3} + \rho_o F_\Lambda = \rho_o \ddot{u}_\Lambda,$$

$$
\begin{aligned}
&L_3U + (\lambda + \mu)\alpha T_2 X_\Gamma(u_{\Gamma,33} - u_{3,\Gamma 3} - u_{R,R\Gamma}) - \\
&\quad - [(2\lambda+\mu)\alpha-\lambda_1]T_2 u_{R,R} - (\mu\alpha - 2\mu_1)T_2 u_{3,3} + \\
&\quad + (2\alpha\beta-\delta_1)T_2\theta + \alpha\beta T_2 X_\Gamma \theta_{,\Gamma} + \rho_0 F_3 = \rho_0 \ddot{u}_3, \\
&(k+\kappa\zeta)\Delta\theta + k_2 T_2 \Delta u_3 + (k_1+k_2)T_2 u_{R,R3} + (\kappa+g)T_2\theta_{,3} - \\
&\quad - (\beta T_0 + \beta\zeta + \delta_1 T_0\zeta)\dot{u}_{R,R} - (a_o T_0 + a_o\zeta + a_1 T_0\zeta)\dot{\theta} - \\
&\quad - \alpha\beta T_2 T_0 X_\Gamma(\dot{u}_{\Gamma,3} - \dot{u}_{3,\Gamma}) = \rho_0 S, \quad (\Lambda,\Gamma = 1,2),
\end{aligned}
\tag{6.4.40}
$$

where $\Delta f = f_{,AA}$, $U = (u_1,u_2,u_3,\theta)$ and

$$
\begin{aligned}
L_A U &= (\mu+\mu_1\zeta)\Delta u_A + [\lambda+\mu+(\lambda_1+\mu_1)\zeta]u_{R,RA} - \\
&\quad - (\beta+\delta_1\zeta)\theta_{,A}.
\end{aligned}
\tag{6.4.41}
$$

The coefficients of the equations (6.4.40) are functions of X_A.

Clearly, if $T_1 = T_2 = 0$ then we recover the displacement-temperature equations of classical thermoelasticity.

It follows from (6.4.40) that in the presence of the initial nonuniform temperature field (6.4.34), the heat equation for isotropic rigid heat conductors is

$$
(k + \kappa\zeta)\Delta\theta + bT_2\theta_{,3} - (c + c_1\zeta)\dot{\theta} = -\rho_0 S, \tag{6.4.42}
$$

where

$$
b = \kappa + g, \quad c = a_o T_0, \; c_1 = a_o + a_1 T_0.
$$

If $S = 0$ and B^* is in equilibrium, then the equation (6.4.42) becomes

$$
(k + \kappa\zeta)\Delta\theta + bT_2\theta_{,3} = 0. \tag{6.4.43}
$$

Assume that $\theta = \theta(X_3)$. When $T_2 = 0$, the solution of the equation (6.4.43) is $\theta^{(1)} = A_1 + A_2X_3$, where A_α are arbitrary constants. If $T_2 \neq 0$, then the solution of the equation (6.4.43) is given by

$$\theta^{(2)} = C_1^o(k + \kappa\zeta)^{-g/\kappa} + C_2^o,$$

where C_α^o are arbitrary constants. The constants C_α^o can be determined, for example, from the boundary conditions $\theta(h_\alpha) = t_\alpha$, if $\kappa \neq 0$ and $g \neq 0$.

If $T_2 = 0$, then the equations (6.4.38) reduce to

$$S_{AB} = \lambda^* u_{R,R}\delta_{AB} + \mu^*(u_{A,B} + u_{B,A}) - \beta^*\theta\delta_{AB},$$
$$\eta = \beta^* u_{R,R} + a^*\theta, \quad \Phi_A = k^*\theta_{,A}, \tag{6.4.44}$$

where

$$\lambda^* = \lambda + \lambda_1 T_1, \; \mu^* = \mu + \mu_1 T_1, \; \beta^* = \beta + \delta_1 T_1, \; a^* = a_o + a_1 T_1,$$
$$k^* = k + \kappa T_1. \tag{6.4.45}$$

We recall that the constitutive coefficients λ, μ, β, a_o, λ_1, μ_1, δ_1, a_1, κ and k are evaluated at the constant absolute temperature T_0. The equations (6.4.44) are the constitutive equations for B when B is obtained from B_0 by a small deformation due to the constant temperature variation T_1. The body B is isotropic and the constitutive coefficients λ^*, μ^*, β^*, a^*, and k^* can be calculated by (6.4.45) if λ, μ, β, a_o, λ_1, μ_1, δ_1, a_1, κ and k are known.

7 Elastic dielectrics

7.1 Basic equations

The interaction of electromagnetic fields with elastic dielectrics has been the subject of many investigations. The equations for infinitesimal dynamical deformations and weak fields superimposed on a finite deformation and strong electromagnetic field have been deduced in the pioneering work of Toupin [351]. The photoelastic effect is only one example of numerous phenomena which are predicted by this theory. The main aim of this part is to establish the basic equations for infinitesimal displacements and weak fields superimposed on finite deformations of an elastic dielectric. Our treatment is based upon the dynamical theory of elastic dielectrics developed by Toupin [351]. This theory is a v^2/c^2 approximation ($\underline{v}$ is the velocity field in the material and c is the velocity of light in vacuum) to a relavistically invariant theory for an elastic dielectric (cf. Grot [144]).

In what follows we denote by R_0 the region occupied by the dielectric at time t_0. Let R be the current configuration of the body. The Maxwell-Lorenz theory of the electromagnetic field in a non-conducting dielectric is based upon the following balance laws relating the electric field $\underline{E}$, the magnetic flux density $\underline{B}$, and the polarization $\underline{P}$:

Faraday's law

$$\int_{\partial S} \underline{E} \cdot d\underline{x} = -\frac{1}{c}\frac{d}{dt}\int_S \underline{B} \cdot d\underline{a}, \tag{7.1.1}$$

Ampère's law

$$\int_{\partial S} \underline{H} \cdot d\underline{x} = \frac{1}{c}\frac{d}{dt}\int_S \underline{D} \cdot d\underline{a}, \tag{7.1.2}$$

Gauss's law

$$\int_{\partial\omega} \underline{D} \cdot d\underline{a} = 0, \tag{7.1.3}$$

conservation of magnetic flux

$$\int_{\partial\omega} \underline{B} \cdot d\underline{a} = 0, \tag{7.1.4}$$

where ω is an arbitrary material volume, S is an arbitrary open material surface, and

$$\underline{\mathcal{E}} = \underline{E} + \frac{1}{c}\,\underline{\dot{x}} \times \underline{B}, \qquad \underline{\mathcal{H}} = \underline{H} - \frac{1}{c}\,\underline{\dot{x}} \times \underline{D}, \quad \underline{D} = \underline{E} + \underline{P},$$
$$\underline{H} = \underline{B} + \frac{1}{c}\,\underline{\dot{x}} \times \underline{P}. \tag{7.1.5}$$

Here $\underline{D}$ is the electric displacement, and $\underline{H}$ is the magnetic field. The balance laws yield the local equations

$$e_{irs}\mathcal{E}_{s,r} + \frac{1}{c}\,\overset{+}{B}_i = 0, \qquad B_{i,i} = 0,$$
$$e_{irs}\mathcal{H}_{s,r} - \frac{1}{c}\,\overset{+}{D}_i = 0, \qquad D_{i,i} = 0, \tag{7.1.6}$$

on $R \times (t_0, t_1)$, where the derivative $(^+)$ is defined as

$$\overset{+}{F}_i = \dot{F}_i + \dot{x}_j F_{i,j} - \dot{x}_{i,j} F_j + \dot{x}_{j,j} F_i.$$

Upon carrying (7.1.5) in (7.1.6) we obtain the equations

$$e_{irs}E_{s,r} + \frac{1}{c}\,\dot{B}_i = 0, \qquad B_{i,i} = 0,$$
$$e_{irs}H_{s,r} - \frac{1}{c}\,\dot{D}_i = 0, \qquad D_{i,i} = 0. \tag{7.1.7}$$

It is useful to have these equations expressed in terms of quantities referred to the configuration R_0. We define $\underline{\hat{E}}$, $\underline{\hat{B}}$, $\underline{\Pi}$, $\underline{\hat{D}}$, $\underline{\hat{H}}$, $\underline{\hat{\mathcal{E}}}$ and $\underline{\hat{\mathcal{H}}}$ by

$$\hat{E}_A = x_{i,A}E_i, \quad B_i = \frac{1}{J}\, x_{i,A}\hat{B}_A, \quad P_i = \frac{1}{J}\, x_{i,A}\Pi_A,$$
$$D_i = \frac{1}{J}\, x_{i,A}\hat{D}_A, \quad \hat{H}_A = x_{i,A}H_i, \quad \hat{\mathcal{E}}_A = x_{i,A}\mathcal{E}_i, \quad \hat{\mathcal{H}}_A = x_{i,A}\mathcal{H}_A. \tag{7.1.8}$$

The balance laws (7.1.1)-(7.1.4) can be written in the form

$$\int_{\partial S} \hat{\underline{E}} \cdot d\underline{X} = -\frac{1}{c}\frac{d}{dt}\int_{S} \hat{\underline{B}} \cdot d\underline{A}, \quad \int_{\partial\Omega} \hat{\underline{B}} \cdot d\underline{A} = 0,$$

$$\int_{\partial S} \hat{\underline{H}} \cdot d\underline{X} = \frac{1}{c}\frac{d}{dt}\int_{S} \hat{\underline{D}} \cdot d\underline{A}, \quad \int_{\partial\Omega} \hat{\underline{D}} \cdot d\underline{A} = 0.$$

Clearly, ω and S are the images by motion of Ω and S, respectively. These balance laws imply that

$$e_{KLM}\hat{E}_{M,L} + \frac{1}{c}\dot{\hat{B}}_K = 0, \qquad \hat{B}_{L,L} = 0,$$
$$e_{KLM}\hat{H}_{M,L} - \frac{1}{c}\dot{\hat{D}}_K = 0, \qquad \hat{D}_{L,L} = 0, \tag{7.1.9}$$

on $R_o \times (t_0, t_1)$. We have

$$\hat{E}_A = \hat{E}_A + \frac{1}{c} e_{AMK}V_M\hat{B}_K, \quad \hat{H}_M = \hat{H}_M - \frac{1}{c} e_{MAB}V_A\hat{D}_B,$$
$$\hat{D}_A = \Pi_A + JC^{(-1)}_{AB}\hat{E}_B, \qquad \hat{H}_A = \frac{1}{J} C_{AM}\hat{B}_M + \frac{1}{c} e_{AKL}V_K\Pi_L, \tag{7.1.10}$$
$$V_A = X_{A,r}\dot{x}_r.$$

In the theory of Toupin [351], the local form of the balance of linear momentum is

$$t_{ji,j} + g_i + \rho f_i = \rho\ddot{x}_i, \tag{7.1.11}$$

where t_{ij} is the Cauchy stress tensor, $\underline{f}$ is the classical body force per unit mass, and $\underline{g}$ is the body force due to the electromagnetic field, defined by

$$\underline{g} = -\underline{E}\,\mathrm{div}\,\underline{P} + \frac{1}{c}(\overset{+}{\underline{P}} \times \underline{B}). \tag{7.1.12}$$

In view of (7.1.12), the equations (7.1.11) become

$$t_{ji,j} - E_iP_{r,r} + \frac{1}{c} e_{ijk}\overset{+}{P}_jB_k + \rho f_i = \rho\ddot{x}_i, \tag{7.1.13}$$

on $R \times (t_0, t_1)$. The balance of angular momentum and the balance of energy are expressed in the forms

$$\frac{d}{dt}\int_{\omega} \rho \underline{x} \times \ddot{\underline{x}}\, dv = \int_{\omega} \underline{x} \times (\rho \underline{f} + \underline{g}) dv + \int_{\partial\omega} \underline{x} \times \underline{t}\, da,$$

$$\frac{d}{dt}\int_{\omega} \rho(\frac{1}{2}\dot{\underline{x}}^2 + \varepsilon) dv = \int_{\omega} [(\rho \underline{f} + \underline{g})\cdot\dot{\underline{x}} + \underline{E}\cdot\overset{+}{\underline{P}}] dv + \int_{\partial\omega} \underline{t}\cdot\dot{\underline{x}}\, da,$$

respectively. The local form of the balance of angular momentum is $t_{ij} = t_{ji}$. From the balance of energy we obtain the equation of energy

$$\rho\dot{\varepsilon} = t_{ji}v_{i,j} + E_i\overset{+}{P}_i, \tag{7.1.14}$$

on $R \times (t_0,t_1)$. The equations (7.1.13) can be written in the form

$$T_{Ai,A} - X_{K,i}\hat{E}_K\Pi_{M,M} + \frac{1}{c} e_{MAL}X_{M,i}\dot{\Pi}_A\dot{\hat{B}}_L + \rho_0 f_i = \rho_0\ddot{x}_i, \tag{7.1.15}$$

on $R_o \times (t_0,t_1)$. In deriving (7.1.15) we have used the relations (1.1.19), (7.1.8) and

$$(\frac{1}{J}x_{i,A})_{,i} = 0,\ \dot{J} = Jv_{i,i},\ P_{i,i} = \frac{1}{J}\Pi_{A,A},\ \overset{+}{P}_i = \frac{1}{J}x_{i,A}\dot{\Pi}_A. \tag{7.1.16}$$

The equation (7.1.14) can be put in the form

$$\rho_0\dot{\varepsilon} = T_{AB}\dot{E}_{AB} + \hat{E}_A\dot{\Pi}_A. \tag{7.1.17}$$

An elastic dielectric is defined as a material for which the following constitutive equations hold at each point $\underline{X}$ and for all time t,

$$\varepsilon = \varepsilon(E_{AB},\Pi_M),\quad T_{KL} = T_{KL}(E_{AB},\Pi_M),\quad \hat{E}_K = \hat{E}_K(E_{AB},\Pi_M). \tag{7.1.18}$$

As in Section 1.1, from (7.1.17) and (7.1.18) we obtain

$$T_{AB} = \frac{\partial e}{\partial E_{AB}},\quad \hat{E}_A = \frac{\partial e}{\partial \Pi_A}, \tag{7.1.19}$$

where $e = \rho_o\varepsilon$.

The jump conditions, which follow in the usual manner from the integral balance laws, must hold at any surface Γ of possible discontinuity in the fields. These conditions are (see, for example, Toupin [351] and Grot [144])

$$[\underline{E} + \frac{1}{c}\,\underline{w} \times \underline{B}] \times \underline{n} = \underline{0}, \quad [\underline{H} - \frac{1}{c}\,\underline{w} \times \underline{D}] \times \underline{n} = \underline{0}, \quad [\underline{B}] \cdot \underline{n} = 0,$$

$$[\underline{D}] \cdot \underline{n} = 0, \quad [(\dot{x}_i - w_i)(\rho\dot{x}_j + \frac{1}{c}e_{jrs}E_rB_s) - \bar{t}_{ij}]n_j = 0, \tag{7.1.20}$$

$$[\rho\varepsilon(\dot{x}_i - w_i) - \bar{t}_{ij}\dot{x}_j]n_i = 0 \text{ on } \Gamma,$$

where $\underline{w}$ is the velocity of the discontinuity surface, $\underline{n}$ is its unit normal, and

$$\bar{t}_{ij} = t_{ij} + E_iE_j + B_iB_j - \frac{1}{2}(E^2 + B^2)\delta_{ij}. \tag{7.1.21}$$

7.2 Equations for infinitesimal deformations and weak fields superimposed on a finite deformation and strong electromagnetic field

We consider three states of the body: the state R_0 and two current configurations, R and R*. We assume that R_0 is a configuration of equilibrium. As before, we call R the primary state and R* the secondary state. The quantities associated with the configuration R^* will be denoted with an asterisk. We assume that the point $\underline{X}$ in the configuration R_0 moves to $\underline{x}$ in R and to $\underline{y}$ in the configuration R^*. Now let

$$u_i = y_i - x_i, \quad \pi_A = \Pi_A^* - \Pi_A, \quad \hat{b}_A = \hat{B}_A^* - \hat{B}_A. \tag{7.2.1}$$

We assume that u_i, π_A, and $\hat{b}_A$ are small, i.e.

$$u_i = \alpha u_i', \quad \pi_A = \alpha\pi_A', \quad \hat{b}_A = \alpha\hat{b}_A', \tag{7.2.2}$$

where α is a constant small enough for squares and higher powers to be neglected, and u_i', π_A' and $\hat{b}_A'$ are independent of α.

In this section we shall derive the equations for the infinitesimal displacements and weak fields superimposed on the deformed configuration R.

If we refer the perturbation to the configuration R_o, then we have

$$u_i = u_i(\underline{X},t),\ \pi_A = \pi_A(\underline{X},t),\ \hat{b}_A = \hat{b}_A(\underline{X},t), \qquad (\underline{X},t) \in R_o \times (t_o,t_1). \tag{7.2.3}$$

In what follows we assume that R is a configuration of equilibrium. It is possible to refer the incremental motion to the configuration R, and for this purpose we introduce the fields $T^{*(1)}_{ij}$, $T^{*(2)}_{ij}$, $\bar{D}^*_i$, $\bar{E}^*_i$, $\bar{B}^*_i$, $\bar{\mathcal{E}}^*_i$, $\bar{\mathcal{H}}^*_i$, $\bar{\Pi}^*_i$ defined on $R \times (t_o,t_1)$, by

$$\begin{gathered} t^*_{ij} = \frac{1}{J'}\, y_{i,s}T^{*(1)}_{sj} = \frac{1}{J'}\, y_{i,s}y_{j,r}T^{*(2)}_{sr}, \\ D^*_i = \frac{1}{J'}\, y_{i,s}\bar{D}^*_s,\quad B^*_i = \frac{1}{J'}\, y_{i,s}\bar{B}^*_s,\quad P^*_i = \frac{1}{J'}\, y_{i,s}\bar{\Pi}^*_s, \\ \bar{E}^*_i = y_{j,i}E^*_j,\ \bar{H}^*_i = y_{j,i}H^*_j\ ,\quad \bar{\mathcal{E}}^*_i = y_{j,i}\mathcal{E}^*_j,\ \bar{\mathcal{H}}^*_i = y_{j,i}\mathcal{H}^*_j, \end{gathered} \tag{7.2.4}$$

where J' is given by (1.2.8) and $f_{,i} = \partial f/\partial x_i$. Now let

$$p_i = \bar{\Pi}^*_i - P_i,\quad b_i = \bar{B}^*_i - B_i. \tag{7.2.5}$$

Clearly, we have

$$p_i = \frac{1}{J}\, x_{i,A}\, \pi_A,\quad b_i = \frac{1}{J}\, x_{i,A}\hat{b}_A. \tag{7.2.6}$$

We note that

$$\begin{gathered} t^*_{ij} = \frac{1}{J^*}\, y_{i,A}y_{j,B}T^*_{AB} = \frac{1}{J^*}\, y_{i,A}T_{Aj},\quad D^*_i = \frac{1}{J^*}\, y_{i,A}\hat{D}^*_A, \\ B^*_i = \frac{1}{J^*}\, y_{i,A}\hat{B}^*_A,\quad P^*_i = \frac{1}{J^*}\, y_{i,A}\Pi^*_A,\ \hat{E}^*_A = y_{i,A}E^*_i, \\ \hat{H}^*_A = y_{i,A}H^*_i,\ \hat{\mathcal{E}}^*_A = y_{i,A}\, \mathcal{E}^*_i,\ \hat{\mathcal{H}}^*_A = y_{i,A}\mathcal{H}^*_i,\quad J^* = JJ', \\ \mathcal{E}^*_i = E^*_i + \frac{1}{c}\, e_{ijk}\dot{y}_jB^*_k,\quad \mathcal{H}^*_i = H^*_i - \frac{1}{c}\, e_{irs}\dot{y}_rP^*_s. \end{gathered} \tag{7.2.7}$$

If we use (7.2.2), (1.2.17) and (1.2.18) then to a second order approximation, we obtain

$$\frac{\partial e^*}{\partial E^*_{KL}} = \frac{\partial e}{\partial E_{KL}} + A_{KLMN} x_{i,M} x_{j,N} e_{ij} + L_{KLA}\, \pi_A,$$
$$\frac{\partial e^*}{\partial \Pi^*_A} = \frac{\partial e}{\partial \Pi_A} + L_{MNA} x_{i,M} x_{j,N} e_{ij} + G_{AB}\, \pi_B, \tag{7.2.8}$$

where

$$A_{KLMN} = \frac{\partial^2 e}{\partial E_{KL}\partial E_{MN}}, \quad L_{MNA} = \frac{\partial^2 e}{\partial E_{MN}\partial \Pi_A}, \quad G_{AB} = \frac{\partial^2 e}{\partial \Pi_A \partial \Pi_B}. \tag{7.2.9}$$

Clearly, we have

$$A_{KLMN} = A_{MNKL} = A_{LKMN}, \quad L_{MNA} = L_{NMA}. \tag{7.2.10}$$

It follows from (7.1.19) and (7.2.8) that

$$T^*_{AB} = T_{AB} + A_{ABMN} x_{i,M} x_{j,N} e_{ij} + L_{ABK} \pi_K, \tag{7.2.11}$$
$$\hat{E}^*_A = \hat{E}_A + L_{MNA} x_{i,M} x_{j,N} e_{ij} + G_{AB} \pi_B.$$

From (1.2.16), (1.2.18), (7.2.10) and (7.2.11) we obtain

$$T^*_{KL} = T_{KL} + A_{KLMN} x_{i,M} u_{i,N} + L_{KLM} \pi_M,$$
$$\hat{E}^*_A = \hat{E}_A + L_{MNA} x_{i,M} u_{i,N} + G_{AB} \pi_B. \tag{7.2.12}$$

α) <u>The equations of incremental motion referred to the configuration</u> R. We now wish to establish the equations for the displacements u_i and the fields p_i and b_i. We assume that

$$u_i = u_i(\underline{x},t), \quad p_i = p_i(\underline{x},t),$$
$$b_i = b_i(\underline{x},t), \quad (\underline{x},t) \in R \times (t_0,t_1). \tag{7.2.13}$$

The electromagnetic equations for R* can be written in the form

$$e_{irs}\bar{\mathcal{E}}^*_{s,r} + \frac{1}{c}\dot{\bar{B}}^*_i = 0, \quad \bar{B}^*_{i,i} = 0,$$
$$e_{irs}\bar{\mathcal{H}}^*_{s,r} - \frac{1}{c}\dot{\bar{D}}^*_i = 0, \quad \bar{D}^*_{i,i} = 0, \qquad (7.2.14)$$

on $R \times (t_0,t_1)$, where

$$\bar{\mathcal{E}}^*_i = \bar{E}^*_i + \frac{1}{c} e_{irs}v^*_r\bar{B}^*_s, \quad \bar{\mathcal{H}}^*_i = \bar{H}^*_i - \frac{1}{c} e_{irs}v^*_r\bar{D}^*_s,$$
$$\bar{D}^*_i = \bar{\Pi}^*_i + J'\gamma^{*(-1)}_{ij}\bar{E}^*_j, \quad \bar{H}^*_i = \frac{1}{J'}\gamma^*_{ij}\bar{B}^*_j + \frac{1}{c} e_{irs}v^*_r\bar{\Pi}^*_s, \qquad (7.2.15)$$
$$v^*_i = \frac{\partial x_i}{\partial y_j}\dot{y}_j, \quad \gamma^*_{ij} = y_{r,i}y_{r,j}.$$

The equations of motion for R* become

$$T^{*(1)}_{ji,j} - \frac{\partial x_k}{\partial y_i}(\bar{\mathcal{E}}^*_k\bar{\Pi}^*_{m,m} - \frac{1}{c} e_{krs}\dot{\bar{\Pi}}^*_r\bar{B}^*_s) + \rho f^*_i = \rho\ddot{y}_i, \qquad (7.2.16)$$

on $R \times (t_0,t_1)$. We introduce the notations

$$\varepsilon_i = \bar{\mathcal{E}}^*_i - \mathcal{E}_i, \; \chi_i = \bar{\mathcal{H}}^*_i - \mathcal{H}_i, \quad d_i = \bar{D}^*_i - D_i,$$
$$h_i = \bar{H}^*_i - H_i, \; e_i = \bar{E}^*_i - E_i, \quad s_{ij} = T^{*(1)}_{ij} - t_{ij}. \qquad (7.2.17)$$

By (7.2.4) and (7.2.7),

$$T^{*(1)}_{ij} = \frac{1}{J} x_{i,L}x_{r,K}T^*_{LK}y_{j,r}, \quad \bar{\mathcal{E}}^*_i = X_{A,i}\hat{\mathcal{E}}^*_A. \qquad (7.2.18)$$

It follows from (1.1.19), (7.2.11) and (7.2.18) that

$$T^{*(1)}_{ij} = t_{ij} + C_{ijrs}e_{rs} + S_{ijr}p_r + t_{ir}u_{j,r},$$
$$\bar{\mathcal{E}}^*_i = \mathcal{E}_i + S_{rsi}e_{rs} + \zeta_{ij}p_j, \qquad (7.2.19)$$

where

$$C_{ijrs} = \frac{1}{J} x_{i,K}x_{j,L}x_{r,M}x_{s,N}A_{KLMN},$$

$$S_{ijr} = x_{i,A}x_{j,B}x_{K,r}L_{ABK}, \qquad (7.2.20)$$

$$\zeta_{ij} = JX_{A,i}X_{B,j}G_{AB}.$$

Clearly,

$$C_{ijrs} = C_{jirs} = C_{rsij}, \; S_{ijr} = S_{jir}, \; \zeta_{ij} = \zeta_{ji}. \qquad (7.2.21)$$

By (7.2.17) and (7.2.19),

$$s_{ij} = C_{ijrs}e_{rs} + S_{ijr}p_r + t_{ir}u_{j,r}, \; \varepsilon_i = S_{rsi}e_{rs} + \zeta_{ij}p_j. \qquad (7.2.22)$$

We note that the stress s_{ij} is not symmetric unless the initial stress vanishes.

Forming differences of corresponding terms in the two sets of equations (7.1.6) and (7.2.14), we obtain

$$e_{irs}\varepsilon_{s,r} + \frac{1}{c}\dot{b}_i = 0, \; b_{i,i} = 0, \; e_{irs}\chi_{s,r} - \frac{1}{c}\dot{d}_i = 0, \; d_{i,i} = 0. \qquad (7.2.23)$$

We recall that R is in equilibrium. It follows from (7.1.5), (7.2.2) and (7.2.15) that

$$\varepsilon_i = e_i + \frac{1}{c} e_{irs}\dot{u}_r B_s, \quad \chi_i = h_i - \frac{1}{c} e_{irs}\dot{u}_r D_s,$$

$$d_i = p_i + e_i - u_{i,j}E_j - u_{j,i}E_j + u_{s,s}E_i, \qquad (7.2.24)$$

$$h_i = b_i + \frac{1}{c} e_{irs}\dot{u}_r P_s + u_{i,j}B_j + u_{j,i}B_j - u_{r,r}B_i.$$

The equations of motion (7.1.13) and (7.2.16) lead to the equations

$$s_{ji,j} - p_{r,r}E_i + (E_m u_{m,i} - \varepsilon_i)P_{r,r} + \frac{1}{c} e_{irs}\dot{p}_r B_s + \rho F_i = \rho\ddot{u}_i \qquad (7.2.25)$$

on $R \times (t_0, t_1)$, where $F_i = f_i^* - f_i$.

Thus, the basic equations for the functions u_i, p_i and b_i consists of the equations (1.2.18), (7.2.22)-(7.2.25).

When the initial stress and the primary fields vanish, then we obtain Voigt's linear theory of piezoelectricity. In this case the equations (7.2.23) and (7.2.24) reduce to

$$e_{irs}e_{s,r} + \frac{1}{c}\dot{b}_i = 0, \quad b_{i,i} = 0,$$
$$e_{irs}b_{s,r} - \frac{1}{c}(\dot{p}_i + \dot{e}_i) = 0, \quad (p_i + e_i)_{,i} = 0. \tag{7.2.26}$$

The equations of motion (7.2.25) become

$$s_{ji,j} + \rho F_i = \rho \ddot{u}_i, \tag{7.2.27}$$

and the constitutive equations simplify to the following set

$$s_{ij} = C_{ijrs}e_{rs} + S_{ijr}p_r,$$
$$e_i = S_{rsi}e_{rs} + \zeta_{ij}p_j. \tag{7.2.28}$$

Equations (7.2.26)-(7.2.28), (1.2.18) are the basic equations of the classical theory of piezoelectricity.

β) <u>The basic equations referred to the configuration</u> R_o. We now establish the equations governing the fields u_i, π_A and $\hat{b}_A$ defined by (7.2.1) and (7.2.3). We introduce the notations

$$\hat{\varepsilon}_A = \hat{E}^*_A - \hat{E}_A, \quad \chi_A = \hat{H}^*_A - \hat{H}_A, \quad \hat{e}_A = \hat{E}^*_A - \hat{E}_A,$$
$$\hat{h}_A = \hat{H}^*_A - \hat{H}_A, \quad d_A = \hat{D}^*_A - \hat{D}_A, \quad S_{Ai} = T^*_{Ai} - T_{Ai}. \tag{7.2.29}$$

It follows from (1.1.19) and (7.2.12) that

$$S_{Ai} = H_{AijN}u_{j,N} + N_{AiK}\pi_K + T_{AN}u_{i,N},$$
$$\hat{\varepsilon}_A = N_{MiA}u_{i,M} + G_{AB}\pi_B, \tag{7.2.30}$$

where

$$H_{AijN} = A_{ABMN}x_{i,B}x_{j,M}, \quad N_{AiK} = L_{ABK}x_{i,B},$$
$$G_{AiK} = R_{AMK}x_{i,M}. \tag{7.2.31}$$

The fields $\hat{\mathcal{E}}^*_A$, $\hat{B}^*_A$, $\hat{\mathcal{H}}^*_A$ and $\hat{D}^*_A$ satisfy the equations

$$e_{KLM}\hat{\mathcal{E}}^*_{M,L} + \frac{1}{c}\dot{\hat{B}}^*_K = 0, \quad \hat{B}^*_{L,L} = 0,$$
$$e_{KLM}\hat{\mathcal{H}}^*_{M,L} - \frac{1}{c}\dot{\hat{D}}^*_K = 0, \quad \hat{D}^*_{L,L} = 0, \tag{7.2.32}$$

on $R_o \times (t_0, t_1)$. We note that

$$\hat{\mathcal{E}}^*_A = \hat{E}^*_A + \frac{1}{c} e_{AMK}V^*_M\hat{B}^*_K, \quad \hat{\mathcal{H}}^*_A = \hat{H}^*_A - \frac{1}{c} e_{AKL}V^*_K\hat{D}^*_L,$$
$$\hat{D}^*_A = \Pi^*_A + J^*C^{*(-1)}_{AB}\hat{E}^*_B, \quad \hat{H}^*_A = \frac{1}{J^*} C^*_{AB}\hat{B}^*_B + \frac{1}{c} e_{AKL}V^*_K\Pi^*_L, \tag{7.2.33}$$
$$\dot{y}_i = y_{i,A}V^*_A, \quad C^*_{AB} = y_{i,A}y_{i,B}.$$

The equations of motion for R^* can be written in the form

$$T^*_{Ai,A} - \frac{\partial X_A}{\partial y_i}\left(\hat{\mathcal{E}}^*_K\Pi^*_{M,M} - \frac{1}{c} e_{KAL}\dot{\Pi}^*_A\hat{B}^*_L\right) + \rho_0 f^*_i = \rho_0\ddot{y}_i, \tag{7.2.34}$$

on $R_o \times (t_0, t_1)$.

Forming differences of corresponding terms in the two sets of equations (7.1.9) and (7.2.32), we obtain

$$e_{KLM}\hat{\varepsilon}_{M,L} + \frac{1}{c}\dot{\hat{b}}_K = 0, \quad \hat{b}_{L,L} = 0,$$
$$e_{KLM}\hat{\chi}_{M,L} - \frac{1}{c}\dot{\hat{d}}_K = 0, \quad \hat{d}_{L,L} = 0, \tag{7.2.35}$$

on $R_o \times (t_0, t_1)$. From (7.1.10) and (7.2.33) we get

$$\hat{\varepsilon}_A = \hat{e}_A + \frac{1}{c} e_{AMK} X_{M,i} \dot{u}_i \hat{B}_K, \quad \hat{\chi}_A = \hat{h}_A - \frac{1}{c} e_{AKL} X_{K,i} \dot{u}_i \hat{D}_L,$$

$$\hat{d}_A = \pi_A + JC_{AB}^{(-1)} \hat{e}_B + J(C_{AB}^{(-1)} \hat{E}_B u_{r,r} - X_{A,i} C_{BM}^{(-1)} \hat{E}_B u_{i,M} -$$

$$- X_{B,i} C_{AL}^{(-1)} \hat{E}_B u_{i,L}), \tag{7.2.36}$$

$$\hat{h}_A = \frac{1}{J} C_{AM} \hat{b}_M + \frac{1}{c} e_{AKL} X_{K,i} \dot{u}_i \Pi_L + \frac{1}{J} (x_{i,A} \hat{B}_M u_{i,M} +$$

$$+ x_{i,M} \hat{B}_M u_{i,A} - C_{AM} \hat{B}_M u_{r,r}).$$

The equations (7.1.15) and (7.2.34) imply

$$S_{Ai,A} - X_{K,i} \hat{E}_K \pi_{M,M} - X_{K,j} \Pi_{M,M} (\delta_{ji} \hat{\varepsilon}_K - X_{L,i} u_{j,L} \hat{E}_K) +$$

$$+ \frac{1}{c} e_{KAL} X_{K,i} \dot{\pi}_A \hat{B}_L + \rho_0 F_i = \rho_0 \ddot{u}_i, \tag{7.2.37}$$

on $R_o \times (t_0, t_1)$. Thus, we conclude that the equations governing the infinitesimal displacements u_i and the weak fields π_A and $\hat{b}_A$ are (7.2.30), (7.2.35)-(7.2.37).

In a similar manner, on using (7.1.20), we can obtain the jump conditions corresponding to the preceding perturbations.

7.3 An alternative form of the constitutive equations

In this section we study the constitutive equations when we choose E_{AB} and $\hat{E}_A$ as independent constitutive variables. We make the Legendre-type transformation

$$\sigma = e - \hat{E}_A \Pi_A, \tag{7.3.1}$$

and assume that σ has the functional form

$$\sigma = \sigma(E_{AB}, \hat{E}_M). \tag{7.3.2}$$

It follows from (7.1.17), (7.1.18), (7.3.1) and (7.3.2) that

$$T_{AB} = \frac{\partial \sigma}{\partial E_{AB}}, \quad \Pi_A = -\frac{\partial \sigma}{\partial \hat{\mathcal{E}}_A}. \tag{7.3.3}$$

In this case, by using the notations (7.2.29), we obtain

$$\begin{aligned} T^*_{KL} &= T_{KL} + a_{KLMN} x_{i,M} u_{i,N} - c_{KLM} \hat{\varepsilon}_M, \\ \Pi^*_A &= \Pi_A + c_{MNA} x_{i,M} u_{i,N} + \chi_{AB} \hat{\varepsilon}_B, \end{aligned} \tag{7.3.4}$$

where

$$\begin{aligned} a_{KLMN} &= \frac{\partial^2 \sigma}{\partial E_{KL}\, \partial E_{MN}}, \quad c_{KLM} = -\frac{\partial^2 \sigma}{\partial E_{KL}\, \partial \hat{E}_M}, \\ \chi_{AB} &= -\frac{\partial^2 \sigma}{\partial \hat{E}_A \partial \hat{E}_B}. \end{aligned} \tag{7.3.5}$$

We conclude that the equations (7.2.30) are replaced by

$$\begin{aligned} S_{Ai} &= c_{AijN} u_{j,N} - s_{AiK} \hat{\varepsilon}_K + T_{AN} u_{i,N}, \\ \pi_A &= s_{NiA} u_{i,N} + \chi_{AB} \hat{\varepsilon}_B, \end{aligned} \tag{7.3.6}$$

where

$$c_{AijN} = a_{ABMN} x_{i,B} x_{j,M}, \quad s_{AiK} = c_{ABK} x_{i,B}.$$

Thus, the equations for the determination of u_i, $\hat{\varepsilon}_A$ and $\hat{b}_K$ on $R_o \times (t_o, t_1)$ are given by (7.2.35)-(7.2.37) and (7.3.6).

7.4 The photoelastic effect

We now assume that the body occupying R_o is isotropic. Then, we have

$$\sigma = \sigma(J_1, J_2, \ldots, J_6), \tag{7.4.1}$$

where the invariants J_s $(s = 1,2,\ldots,6)$ are given by

$$J_1 = E_{AA},\ J_2 = E_{AB}E_{AB},\ J_3 = E_{AK}E_{KL}E_{LB},\ J_4 = \hat{E}_A\hat{E}_A,$$
$$J_5 = E_{AB}\hat{E}_A\hat{E}_B, \qquad J_6 = E_{AK}E_{KL}\hat{E}_A\hat{E}_L. \tag{7.4.2}$$

Let us study the polarizability tensor χ_{AB} defined by (7.3.5). In view of (7.4.2), we have

$$\frac{\partial J_4}{\partial \hat{E}_A} = 2\hat{E}_A,\ \frac{\partial J_5}{\partial \hat{E}_A} = 2E_{AL}\hat{E}_L,\ \frac{\partial J_6}{\partial \hat{E}_A} = 2E_{AL}E_{LM}\hat{E}_M. \tag{7.4.3}$$

It follows from (7.3.5), (7.4.1) and (7.4.3) that

$$\begin{aligned}\chi_{AB} = & - 2(\frac{\partial\sigma}{\partial J_1}\delta_{AB} + \frac{\partial\sigma}{\partial J_5}E_{AB} + \frac{\partial\sigma}{\partial J_6}E_{AL}E_{LB}) - \\ & - 4[\frac{\partial^2\sigma}{\partial J_4^2}\hat{E}_A\hat{E}_B + \frac{\partial^2\sigma}{\partial J_4\partial J_5}(E_{AL}\hat{E}_B + E_{BL}\hat{E}_A)\hat{E}_L + \\ & + \frac{\partial^2\sigma}{\partial J_4\partial J_6}(E_{AL}\hat{E}_B + E_{BL}\hat{E}_A)E_{LM}\hat{E}_M + \\ & + \frac{\partial^2\sigma}{\partial J_5^2}E_{AL}\hat{E}_L E_{BM}\hat{E}_M + \frac{\partial^2\sigma}{\partial J_5\partial J_6}(E_{AK}E_{BL} + \\ & + E_{BK}E_{AL})E_{KN}\hat{E}_N\hat{E}_L + \frac{\partial^2\sigma}{\partial J_6^2}E_{AL}E_{LM}\hat{E}_M\ E_{BN}E_{NK}\hat{E}_K].\end{aligned} \tag{7.4.4}$$

We now assume that in the primary state we have $\hat{E}_A = 0$. Then, the polarizability tensor has the form

$$\chi_{AB} = A_o\ \delta_{AB} + A_1E_{AB} + A_2E_{AL}E_{LB}, \tag{7.4.5}$$

where A_o, A_1 and A_2 are functions of the strain invariants J_1, J_2 and J_3. From (7.4.5), we see that the polarizability tensor in the deformed dielectric has the same principal directions as the strain tensor E_{AB}. Because the dielectric tensor $\chi_{AB} + \delta_{AB}$ will have, in general, unequal principal values, the material will be birefringent, i.e. exhibit the photoelastic effect.

7.5 Bibliographical notes

The results presented in this chapter are intended only to illustrate the relevance and applicability of the theory of initially stressed bodies to elastic dielectrics. In the following, an attempt is made to give a brief review of the problems investigated by various authors. Verma and Chaudhry [364] and Croitoro [88] have studied the small twist superimposed on large uniform extensions of a hollow cylindrical elastic dielectric. The effect of initial stress in piezoelectric plates has been investigated by Lee, Wang and Markenscoff [242]. Kaliski and Nowacki, in a long series of publications (see [215]), have treated the problem of waves in elastic and thermoelastic bodies with electromagnetic effects included. Dunkin and Eringen [109] have studied the plane waves in an infinite space with finite conductivity in uniform magnetic and electric fields, and investigated the vibration of an infinite plate in a strong uniform magnetic field. The small torsional vibrations of an elastic, conducting circular cylinder subject to an initial, strong, uniform axial magnetic field have been investigated by Şuhubi [335]. Yu and Tang [386] have discussed the propagation of plane infinitesimal harmonic waves in a prestressed perfect electric conductor. The propagation and growth of plane acceleration discontinuities propagating into a homogeneously deformed elastic dielectric have been studied by McCarthy [253], [255] and McCarthy and Green [254]. One dimensional shock waves in elastic dielectrics have been studied by Chen and McCarthy [75]. A continuous dependence result for one-dimensional dielectrics has been given by Craine [87]. In [331], Soós has constructed some continuous generalized models for describing nonlinear and nonlocal behaviour of a system formed by an elastic solid with microstructure, electrically polarizable and by an electro-magnetic Maxwellian field. The theory of small displacements and weak electromagnetic fields superimposed on large deformations and strong electromagnetic fields is also studied. The general theory of electromagnetic thermoelastic solids has been considered in various works (see Jordan and Eringen [211], Tiersten [346], Parkus [288], Grot [144], Chowdhury and Glockner [80], Hsieh [183]). Willson [380] has studied the propagation of a plane magneto-thermoelastic wave into a homogeneous and isotropic medium in the presence of a magnetic field. Magneto-thermoelastic acceleration waves in a prestressed body have been investigated by McCarthy [255]. The propagation of surfaces waves in an initially stressed magneto-thermoelastic medium has been discussed by Chandrasekharaiah [65]. Iannéce [184] and Beevers and Craine [26] have established various results concerning the stability of thermoelastic dielectrics.

Our guideline in compiling the bibliography has been to select those works which directly bear on our treatment and provide a source for further related references. Completeness was neither achieved nor intended.

References

[1] R. Abeyaratne, C.O. Horgan and D.T. Chung, Saint-Venant end effects for incremental plane deformations of incompressible nonlinearly elastic materials. J. Appl. Mech. 52 (1985), 847-852.

[2] R.A. Adams, Sobolev Spaces. Academic Press, New York, 1975.

[3] V.M. Aleksandrov and N.K. Arutyunyan, Contact problems for prestressed deformed bodies (Russian). Prikl. Mekh. 20 (1984), No. 3, 9-16.

[4] N.A. Alfutov and L.I. Balabuh, On the possibility of solving plate stability problems without a preliminary determination of the initial state of stress (Russian). Prikl.Mat.Meh. 31 (1967), 716-722.

[5] G. Amendola and T. Manacorda, Piccole deformazioni termoelastiche sovraposte ad una deformazione termostatica finita. Riv.Mat.Univ. Parma (4), 5 (1979), 487-501.

[6] I.V. Anan'ev, V.V. Kalinchuk and I.B. Polyakova, On wave excitation by a vibrating stamp in a medium with inhomogeneous initial stresses (Russian). Prikl.Mat.Mekh. 47 (1983), 483-489.

[7] M. Anliker and B.A. Troesch, Lateral vibrations of pretwisted rods with various boundary conditions. ZAMP 14 (1963), 218-236.

[8] C.B. Archambeau, The theory of stress wave radiation from explosions in prestressed media. Geophys. J. Roy. Astron. Soc. 29 (1972), 329-366.

[9] A.E. Armenákas and G. Hermann, Vibrations of infinitely long cylindrical shells under initial stress. AIAA Journal, 1 (1963), 100-106.

[10] J.C. Arya and M. Singh, Controllable infinitesimal deformations superposed on homogeneously deformed compressible elastic dielectrics. Utilitas Math. 6 (1974), 45-60.

[11] S. Yu. Babich, A.N. Guz' and Y.P. Zhuk, Surface waves, waves of Stonely and Love in bodies with initial stresses (Russian). Continuum mechanics. Mater. All-Union Conf., Collect.Rep., Tashkent 1979, 81-102 (1982).

[12] S. Yu. Babich, On contact problems for a prestressed half-plane with regard to friction forces (Ukrainian). Dopov.Akad.Nauk Ukr.RSR, Ser.A 12 (1980), 21-24.

[13] S. Yu. Babich, Axisymmetrical contact problem for bodies with initial stresses (Russian). Dokl.Akad.Nauk Ukr.SSR, Ser.A 6 (1982), 32-35.

[14] S. Yu. Babich, Contact problem of the theory of elasticity for a layer with initial stresses (Russian). Prikl.Mekh. 20 (1984), Nr. 12, 34-40.

[15] S. Yu. Babich and A.N. Guz', General spatial static contact problem for a prestressed elastic half-space (Russian). Prikl.Mat.Mekh. 49 (1985), 438-444.

[16] S. Yu. Babich, Axisymmetric contact problem for a layer on an elastic half-space with initial stress. Sov.Appl.Mech. 21, 1048-1052 (1982); translation from Prikl.Mekh., Kiev 21 (1985), No. 11, 32-36.

[17] S. Yu. Babich, Contact problem for a prestressed bar lying on elastic foundation with initial stresses. Sov. Appl. Mech. 22 (1986), 207-211.

[18] M. Baker and J.L. Ericksen, Inequalities restricting the form of the stress-deformation relations for isotropic elastic solids and Reiner-Rivlin fluids. J.Wash.Acad.Sci. 44 (1954), 33-35.

[19] M.O. Basheleishvili, Solution of plane boundary-value problems of statics of anisotropic elastic solids (Russian). Trudy Vychisl. Tsentra A.N. Gruz. SSR, 3 (1961), 91-105.

[20] Z.P. Bažant, A correlation study of formulations of incremental deformation and stability of continuous bodies. J.Appl.Mech. 38 (1971), 919-928.

[21] M.F. Beatty, Some static and dynamic implications of the general theory of elastic stability. Arch. Rational Mech.Anal. 19 (1965), 167-188.

[22] M.F. Beatty, A theory of elastic stability for incompressible, hyper-elastic bodies. Int.J. Solids Structures 3 (1967), 23-37.

[23] M.F. Beatty, Some theorems in the theory of small deformations superimposed on a finite deformation of a hyperelastic material of grade 2. Int.J. Engng.Sci. 12 (1976), 339-351.

[24] C.E. Beevers, Some continuous dependence results in the linear dynamic theory of anisotropic viscoelasticity. J. Mécanique 14 (1975), 639-651.

[25] C.E. Beevers, Uniqueness and stability in linear viscoelasticity. ZAMP 26 (1975), 177-186.

[26] C.E. Beevers and R.E. Craine, On the thermodynamics and stability of deformable dielectrics. Appl.Sci.Res. 37 (1981), 241-256.

[27] J.A. Belward, Elastic waves in a prestressed Mooney material. Bull. Austral.Math.Soc. 7 (1972), 135-160.

[28] J.A. Belward, Some dynamic properties of a prestressed incompressible hyperelastic material. Bull.Austral.Math.Soc. 8 (1973), 61-73.

[29] J.A. Belward, The propagation of small amplitude waves in prestressed incompressible elastic cylinders. Int.J.Engng.Sci. 14 (1976), 647-659.

[30] Y. Benveniste, One-dimensional wave propagation in an initially deformed incompressible medium with different behaviour in tension and compression. Int.J.Engng.Sci. 19 (1981), 697-711.

[31] J.T. Bergen, D.C. Messersmith and R.S. Rivlin, Stress relaxation for biaxial deformation of filled high polymers. J.Appl.Polymer.Sci. 3 (1960), 153-167.

[32] S. Bhagavantam and E.V. Chelam, Elastic behavior of matter under very high pressures. Uniform compression. Proc.Indian Acad.Sci. 52 (1960), 1-19.

[33] S. Bhagavantam and E.V. Chelam, Elastic behavior of matter under very high pressures. General deformation. J.Indian Inst.Sci. 42 (1960), 29-40.

[34] M.A. Biot, Internal buckling under initial stress in finite elasticity. Proc. Roy.Soc.London, Sér. A 273 (1963), 306-328.

[35] M.A. Biot, Continuum theory of stability of an embedded layer in finite elasticity under initial stress. Quart.J.Math.Appl.Math. 17 (1964), 17-22.

[36] M.A. Biot, Mechanics of Incremental Deformations. New York, Wiley, 1965.

[37] G.A. Bliss, Algebraic Functions, AMS Colloq.Pub., vol.XVI, New York, 1933.

[38] F. Bloom, On continuous dependence for a class of dynamical problems in linear thermoelasticity. ZAMP. 26 (1975), 569-579.

[39] F.E. Browder, Strongly elliptic systems of differential equations. Ann.Math.Studies 33 (1954), 15-51.

[40] L. Brun, Sur l'unicité en thermoélasticité dynamique et diverses expressions analogues à la formule de Clapeyron. C.R.Acad.Sci. Paris 261 (1965), 2584-2587.

[41] L. Brun, Méthodes énergétiques dans les systèmes évolutifs linéaires. Premier Partie: Séparation des énergies, Deuxième Partie: Théorèmes d'unicité. J.Mécanique 8 (1969), 125-166.

[42] E.J. Brunelle, Stability and vibration of transversely isotropic beams under initial stress. J.Appl.Mech. 39 (1972), 819-821.

[43] E.J. Brunelle and R.S. Robertson, Vibrations of an initially stressed thick plate. J.Sound Vibration 45 (1976), 405-416.

[44] T.V. Burchuladze, Application of singular integral equations for solving some boundary-value problem (Russian). Trudy Tbiliskogo Mat. Inst. 28 (1962), 21-31.

[45] B.O. Calleb, Existence of solutions for the prestressed non-linear orthotropic heated plates. Arch.Mech. 29 (1977), 715-722.

[46] G. Caricato, Il teorema di Menabrea per transformazioni non isoterme di un corpo elastico vincolato anisotropo e non omogeneo, con stress iniziale. Atti Accad.Naz.Lincei, Rend., Cl.Sci.fis.mat.natur. VIII, 44 (1968), I: 191-200, II: 363-369.

[47] D.E. Carlson, Linear Thermoelasticity. pp. 297-346 of Flügge's Handbuch der Physik, vol. VI a/2 (Edited by C. Truesdell) Springer-Verlag, Berlin-Heidelberg-New York, 1973.

[48] J. Casey and P.M. Naghdi, An invariant infinitesimal theory of motions superposed on a given motion. Arch.Rational Mech.Anal. 76 (1981), 355-391.

[49] J. Casey and P.M. Naghdi, Small deformations superposed on large deformations of an elastic-plastic material. Int.J.Solids Structures 19 (1983), 1115-1146.

[50] A.-L.Cauchy, Sur l'équilibre et le mouvement intérieur des corps considérés comme des masses continue. Ex.de Math. 4 (1829), 293-319.

[51] H.M. Cekirge and E.S. Şuhubi, Propagation of a plane wave in an initially stressed thermoelastic medium. Bull.Techn.Univ. Istanbul 27 (1974), 98-107.

[52] P.Chadwick, Thermoelasticity. The Dynamical Theory. In Progress in Solid Mechanics. (Edited by I.N. Sneddon and R. Hill), vol. 1, pp. 263-328, North Holland Publ.Co., Amsterdam, 1960.

[53] P. Chadwick and D.W. Windle, Propagation of Rayleigh waves along isothermal and insulated boundaries. Proc.R.Soc.London Ser. A, 280 (1964), 47-71.

[54] P. Chadwick and L.T.C. Seet, Second-order thermoelasticity theory for isotropic and transversely isotropic materials. Trends in Elasticity and Thermoelasticity, Witold Nowacki Anniversery Volume, Wolters-Noordhoff Publ., Groningen, 1971, 29-57.

[55] P. Chadwick and R.W. Ogden, On the definition of elastic moduli. Arch. Rational Mech.Anal. 44 (1971), 41-53.

[56] P. Chadwick and P.K. Currie, The propagation and growth of acceleration waves in heat-conducting elastic materials. Arch. Rational Mech.Anal. 49 (1972), 137-158.

[57] P. Chadwick, Interchange of modal properties in the propagation of harmonic waves in heat-conducting elastic materials. Bull. Austr. Math.Soc. 8 (1973), 75-92.

[58] P. Chadwick and P.K. Currie, On the propagation of generalized transverse waves in heat-conducting elastic materials. J. Elasticity 4 (1974), 301-315.

[59] P. Chadwick and D.A. Jarvis, Interfacial waves in a pre-strained neo-Hookean body. II. Triaxial states of strain. Q.J. Mech. Appl. Math. 32 (1979), 401-418.

[60] P. Chadwick, Basic properties of plane harmonic waves in a prestressed heat-conducting elastic material. J.Therm.Stresses 2 (1979), 193-214.

[61] P. Chadwick and D.A. Jarvis, Surface waves in a prestressed elastic body. Proc.R.Soc.Lond.Ser. A 366 (1979), 517-536.

[62] P. Chadwick and C.F.M. Creasy, Weak nonlinear waves in homogeneously deformed heat-conducting elastic materials. Quart.J.Mech.Appl.Math. 32 (1979), 419-436.

[63] S.K. Chakraborty and S.Dey, The disturbance due to plane and line sources in a prestressed semi-infinite elastic solid. I. Int.J. Solids Structures 18 (1982), 1153-1164.

[64] M. Chakraborty, Distrubance of SH-type due to body forces and due to shearing stress-discontinuity in a pre-stressed semi-fininite viscoelastic medium. Indian J. Pure Appl.Math. 16 (1985), 309-322.

[65] D.S. Chandrasekharaiah, Surface waves in an initially stressed magneto-thermo-elastic medium. Tensor 27 (1973), 21-27.

[66] A. Chattopadhyay and R.K. De, Propagation of Love-type waves in a visco-elastic initially stressed layer overlying a visco-elastic half-space with irregular interface. Rev.Roum.Sci.Tech., Sér.Méc.Appl. 26 (1981), 449-460.

[67] A. Chattopadhyay and A. Keshri, Generation of G-type seismic waves under initial stress. Indian J.Pure Appl.Math. 17 (1986), 1042-1055.

[68] E.V. Chelam, Elastic behavior of matter under very high pressures. Considerations of stability. J.Indian Inst.Sci. 42 (1960), 41, 101-107.

[69] L.W. Chen and J.L. Doong, Large amplitude vibration of an initially stressed thick circular plate. AIAA J. 21 (1983), 1317-1324.

[70] J.K. Chen and C.T. Sun, Nonlinear transient responses of initially stressed composite plates. Comput.Struct. 21 (1985), 513-520.

[71] P.J. Chen, The growth of acceleration waves of arbitrary form in homogeneously deformed elastic materials. Arch. Rational Mech.Anal. 30 (1968), 81-89.

[72] P.J. Chen, On the growth of longitudinal waves in anisotropic elastic materials. Arch.Rational Mech.Anal. 36 (1970), 381-389.

[73] P.J. Chen, On the growth of transverse waves in anisotropic elastic materials. ZAMP 21 (1970), 846-850.

[74] P.J. Chen, Growth and Decay of Waves in Solids, pp. 303-400 of Flügge's Handbuch der Physik, vol. VI a/3, (Edited by C. Truesdell) Springer-Verlag, Berlin-Heidelberg-New York, 1973.

[75] P.J. Chen and M.F. McCarthy, One dimensional shock waves in elastic dielectrics. Ist. Lombardo Accad.Sci.Lett., Rend.Sez. A 107 (1973), 715-727 (1974).

[76] P.J. Chen, M.F. McCarthy and R.T. O'Leary, One-dimensional shock and acceleration waves in deformable dielectric materials with memory. Arch.Rational Mech.Anal. 62 (1976), 189-207.

[77] S. Chiriţa, Uniqueness and continuous dependence results for the incremental thermoelasticity. J.Thermal Stresses 5 (1982), 161-172.

[78] S. Chiriţa and Gh. Rusu, On existence and uniqueness in thermo-elasticity of non-simple materials. Rev. Roum.Sci.Techn., Méc. Appl. 30 (1985), 21-30.

[79] S. Chiriţa, Hölder continuuos dependence and continuous results for incremental elastodynamics, Int.J.Engng.Sci. 24 (1986), 803-812.

[80] K.L. Chowdhury and P.G. Glockner, On thermoelastic dielectrics, Int. J. Solids Structures 13 (1977), 1173-1182.

[81] P. Chowdhury and B. Chakravorti, Stresses around a crack in a semi-infinite initially stressed body of incompressible material. Acta Cienc. Indica 11 (1985), 225-232.

[82] B.D. Coleman and W. Noll, The thermodynamics of elastic materials with heat conduction and viscosity. Arch.Rational Mech.Anal. 13 (1963), 167-178.

[83] B.D. Coleman and V.J. Mizel, Existence of caloric equations of state in thermodynamics. J.Chem.Phys. 40 (1964), 1116-1125.

[84] B.D. Coleman and M.E. Gurtin, Waves in materials with memory. IV. Thermodynamics and the velocity of general acceleration waves. Arch. Rational Mech.Anal. 19 (1965), 317-338.

[85] B.D. Coleman and E.H. Dill, Thermodynamic restrictions on the constitutive equations of electromagnetic theory. ZAMP 22 (1971), 691-702.

[86] B. Collet, One-dimensional acceleration waves in deformable dielectrics with polarization gradients. Int.J.Engng.Sci. 19 (1981), 389-407.

[87] R.E. Craine, A continuous dependence result for one-dimensional nonlinear dielectrics. Int.J.Engng.Sci. 25 (1987), 755-758.

[88] E.M. Croitoro, Perturbational torsion about a finite extension of an elastic dielectric. ZAMP 38 (1987), 450-458.

[89] P. Currie, Controllable infinitesimal deformations in homogeneously deformed compressible elastic materials. Appl.Sci.Research 23 (1970), 212-220

[90] P.K. Currie, Shock waves in homogeneously-strained incompressible elastic materials: The Mooney-Rivlin material. Acta Mech. 14 (1972), 53-58.

[91] C.M. Dafermos, On the existence and the asymptotic stability of solutions to the equations of linear thermoelasticity. Arch.Rational Mech.Anal. 29 (1968), 241-271.

[92] C.M. Dafermos, The second law of thermodynamics and stability. Arch. Rational Mech.Anal. 70 (1979), 167-179.

[93] S.C. Das and S. Dey, Edge waves under initial stress. Appl.Sci. Research 22 (1970), 382-389.

[94] K. Datta, Note on small oscillations of strongly twisted cylindrical bodies. Bull. Calcutta Math.Soc. 60 (1968), 63-70.

[95] W.A. Day, Generalized torsion: The solution of a problem of Truesdell's. Arch.Rational Mech.Anal. 76 (1981), 283-288.

[96] H. Demiray and E.S. Şuhubi, Small torsional oscillations of an initially twisted circular rubber cylinder. Int.J.Engng.Sci. 8 (1970), 19-30.

[97] H. Demiray, Small flexural oscillations of an initially stretched circular cylinder. Int.J.Non-Linear Mech. 6 (1971), 135-141.

[98] H. Demiray, Wave propagation in a prestressed reinforced composite. Rozprawy inz. 24 (1976), 549-558.

[99] S. Dey and S. K. Addy, Reflection and refraction of plane waves under initial stresses at an interface. Int.J.Non-Linear Mech. 14 (1979), 101-110.

[100] S. Dey and S.P. Mukherjee, Propagation, reflection and transmission of waves under initial shear stresses. Int.J.Non-Linear Mech. 18 (1983), 269-277.

[101] S. Dey and S.P. Mukherjee, SH waves in composite media under initial stresses. Rev.Roum.Sci.Techn., Sér.Méc. Appl. 28 (1983), 387-393.

[102] S. Dey and P. Pal Roy, On the effect of initial stresses in two dimensional finite element stress analysis. Indian J. Pure Appl. Math. 15 (1984), 899-926.

[103] S. Dey and M. Chakraborty, Rayleigh waves in granular medium over an initially stressed elastic half space. Rev.Roum.Sci.Techn. Sér.Méc. Appl. 29 (1984), 271-285.

[104] S. Dey, S.K. Chakraborty and M. Chakraborty, Dynamic response of normal moving load in an initially stressed transversely isotropic half-space. Int.J.Solids Struct. 22 (1986), 283-291.

[105] R.S. Dhaliwal, B.M. Singh and J.G. Rokne, Axisymmetric contact and crack problems for an initially stressed neo-Hookean elastic layer. Int.J.Eng.Sci. 18 (1980), 169-179.

[106] R.C. Dixon and A.C. Eringen, A dynamical theory of polar elastic dielectrics. I, II. Int.J.Engng.Sci. 3 (1965), 359-398.

[107] M.C. Dökmeci, Recent advances. Vibrations of piezoelectric crystals. Int.J.Engng.Sci. 18 (1980), 431-448.

[108] D.S. Dugdale, Measurement of internal stress in discs. Int.J.Engng. Sci., 1 (1963), 383-389.

[109] J.W. Dunkin and A.C. Eringen, On the propagation of waves in an electromagnetic elastic solid. Int.J.Engng.Sci. 1 (1963), 461-495.

[110] G. Eason, Wave propagation in a straight elastic rod subjected to initial finite extension and twist. Arch.Mech. 33 (1981), 541-563.

[111] A.H. England and A.E. Green, Steady-state thermoelasticity for initially stressed bodies. Phil.Trans.Roy.Soc.London, Ser. A 253 (1961), 517-542.

[112] J.L. Ericksen and R.A. Toupin, Implications of Hadamard's condition for elastic stability with respect to uniqueness theorems. Canad. J.Math. 8 (1956), 432-436.

[113] J.L. Ericksen, A thermo-kinetic view of elastic stability theory. Int.J.Solids Structures 2 (1966), 573-580.

[114] A.C. Eringen, On the foundations of electroelastostatics. Int. J. Engng.Sci. 1 (1963), 127-153.

[115] A.C. Eringen, Theory of nonlocal electromagnetic elastic solids, J.Math.Phys. 14 (1973), 733-740.

[116] A.C. Eringen and E.S. Şuhubi, Elastodynamics, vol. I. New York, Academic Press, 1974.

[117] A.C. Eringen, Balance Laws. In vol. II of Continuum Physics (Edited by A.C. Eringen), New York, Academic Press, 1975.

[118] A. Falqués, Thermoelasticity and heat conduction with memory effects. J. Thermal Stresses 5 (1982), 145-160.

[119] M. Farshad, On incremental deformations of initially stressed media with couple stresses. J. Franklin Inst. 305 (1978), 125-135.

[120] G. Fichera, Existence Theorems in Elasticity, pp. 347-388 of Flügge's Handbuch der Physik, vol. VI a/2 (Edited by C. Truesdell). Springer-Verlag, Berlin-Heidelberg-New York, 1972.

[121] A.D. Fine and R.T. Shield, Second-order effects in the propagation of elastic waves. Int.J.Solids Structures 2 (1966), 605-620.

[122] J.N. Flavin and A.E. Green, Plane thermoelastic waves in an initially stressed medium. J.Mech.Phys.Solids 9 (1961), 179.

[123] J.N. Flavin, Thermoelastic Rayleigh waves in a prestressed medium. Proc. Cambridge Phil.Soc. 58 (1962), 532-538.

[124] J.N. Flavin, Surface waves in prestressed Mooney material. Quart. J.Mech.Appl.Math. 16 (1963), 441-449.

[125] R.L. Fosdick and R.T. Shield, Small bending of a circular bar superposed on finite extension or compression. Arch. Rational Mech. Anal. 12 (1963), 223-248.

[126] F. Franchi, Wave propagation in heat conducting dielectric solids with thermal relaxation and temperature dependent electric permitivity. Riv.Mat.Univ.Parma, IV, Ser. 11 (1985), 443-461.

[127] R. Frisch-Fay, Buckling of pre-twisted bars. Int.J.Mech.Sci. 15 (1973), 171-181.

[128] G.P. Galdi and S. Rionero, Continuous data dependence in linear elastodynamics on unbounded domains without definiteness conditions on the elasticities. Proc.Roy.Soc.Edinburgh 93 A(1983), 299-306.
[129] G. Galdi and S. Rionero, Weighted energy methods in fluid dynamics and elasticity. Lecture Notes in Mathematics, 1134, Springer, Berlin-Heidelberg-New York, 1985.
[130] J.P. Gallagher and J.N. Flavin, The fundamental integral for small deformations superposed on finite triaxial extension of a neo-Hookean material. ZAMP 20 (1969), 882-890.
[131] R.R. Goldberg, Fourier Transforms. Cambridge Tracts in Mathematics and Mathematical Physics, 1961.
[132] A.E. Green and R.T. Shield, Finite extension and torsion of cylinders. Phil.Trans.Roy.Soc.London A 224 (1951), 47-86.
[133] A.E. Green, R.S. Rivlin and R.T. Shield, General theory of small elastic deformations superposed on finite elastic deformations. Proc.Roy.Soc.London, Ser. A 211 (1952), 128-154.
[134] A.E. Green and A.J.M. Spencer, The stability of a circular cylinder under finite extension and torsion. J.Math.Phys. 37 (1959), 316-338.
[135] A.E. Green and J.E. Adkins, Large Elastic Deformations and Non-Linear Continuum Mechanics. Oxford, Clarendon Press, 1960.
[136] A.E. Green, Torsional vibrations of an initially stressed circular cylinder. In "Problems of Continuum Mechanics" (Muskhelishvili Anniversary Volume), SIAM, Philadelphia, Pennsylvania, 1961, pp. 148-154.
[137] A.E. Green, Thermoelastic stresses in initially stressed bodies. Proc.Roy.Soc.London, Ser. A 266 (1962), 1-19.
[138] A.E. Green, A note on wave propagation in initially deformed bodies. J.Mech.Phys.Solids 11 (1963), 119-126.
[139] A.E. Green and R.S. Rivlin, Multipoar continuum mechanics. Arch. Rational Mech.Anal. 17 (1964), 113-147.
[140] A.E. Green and W. Zerna, Theoretical Elasticity, 2nd Edition. Clarendon Press, Oxford, 1968.
[141] A.E. Green, R.J. Knops and N. Laws, Large deformations, superposed small deformations and stability of elastic rods. Int.J.Solids Structures 4 (1968), 555-577.

[142] A.E. Green and J.Z. Mkrtichian, Elastic solids with different moduli in tension and compression. J.Elasticity 7 (1977), 369-386.

[143] W.A. Green, The growth of plane discontinuities propagating into a homogeneously deformed elastic material. Arch. Rational Mech. Anal. 16 (1964), 79-88; corrections and additional results ibid. 17 (1965), 20-23.

[144] R.A. Grot, Relativistic Continuum Physics: Electromagnetic Interactions. In vol. III of the Continuum Physics (Edited by A.C. Eringen), New York, Academic Press, 1976.

[145] Z.H. Guo, The equations of motion of a circular plate subject to initial strain. Bull.Acad.Polon.Sci. Sér.Sci.Techn. 10 (1962), 63-70.

[146] Z.H. Guo, Equations of small motion of a cylinder subject to a large deformation. Its natural vibration and stability. Bull. Acad.Polon. Sci. Sér.Sci.Techn. 10 (1962), 177-182.

[147] Z.H. Guo, Displacement equations of isentropic motion of a body subject to a finite isothermal initial strain. Bull.Acad.Polon.Sci. Sér.Sci.Techn. 10 (1962), 479-483.

[148] Z.H. Guo, Vibration and stability of a cylinder subject to finite deformation. Arch.Mech.Stosow. 14 (1962), 757-768.

[149] Z.H. Guo, Certain problems of initially deformed plate. Arch.Mech. Stosow. 14 (1962), 779-788.

[150] Z.H. Guo and R. Solecki, Free and forced finite amplitude oscillations of an elastic thick-walled hollow sphere made of incompressible material. Arch.Mech.Stosow 15 (1963), 427-433.

[151] Z.H. Guo and W. Urbanowski, Certain stationary conditions in variated states of finite strain. BullAcad.Polon.Sci. Sér.Sci.Techn. 11 (1963), 27-32.

[152] M.E. Gurtin, Variational principles for linear elastodynamics. Arch.Rational Mech.Anal. 16 (1964), 34-50.

[153] M.E. Gurtin, The Linear Theory of Elasticity, pp. 1-295 of Flügge's Handbuch der Physik, vol. VI a/2 (Edited by C. Truesdell), Springer-Verlag, Berlin-Heidelberg-New York, 1972.

[154] A.N. Guz', On the bifurcation of the equilibrium state of a three-dimensional isotropic elastic solid under large subcritical deformations (Russian). Prikl.Mat.Mech. 34 (1970), 1113-1125.

[155] A.N. Guz', On representation of general solutions of the linearized theory of elasticity of compressible systems (Ukrainian). Dopovidi Akad.Nauk Ukrain.SSR, Ser. A 1975, 700-703.

[156] A.N. Guz', On determination of reduced elastic constants of composite stratified materials with initial stresses (Ukrainian), Dopovidi Akad. Nauk Ukrain.SSR, Ser. A, 1975, 216-219.

[157] A.N. Guz', On the Love waves in bodies with initial strains. (Ukrainian). Dopov.Akad.Nauk Ukrain.SSR, Ser. A (1978), 1092-1095.

[158] A.N. Guz', Aerohydroelasticity problems for bodies with initial stresses (Russian). Prikl.Mekh. 16 (1980), No. 3, 3-21.

[159] A.N. Guz', Complex potentials of the planar linearized problem of elasticity theory (Russian). Prikl.Mekh. 16 (1980), No. 9, 83-97.

[160] A.N. Guz' and S.Yu. Babich, The axisymmetric contact problem for an elastic layer with initial stresses. (Russian). Dokl.Akad.Nauk. SSSR 273 (1983), 1329-1332.

[161] A.N. Guz', Brittle fracture mechanics of initially stressed materials. (Russian). Akad.Nauk.Ukrainskoj SSR. Inst.Mekh. "Naukova Dumka", Kiev, 1983.

[162] A.N. Guz' and V.B.Rudnitskij, Contact problem on the pressure of an elastic die against an elastic half-space with initial stresses (Russian). Prikl.Mekh. 20 (1984), No. 8, 3-11.

[163] A.N. Guz', N.A. Sitenok and A.P. Zhuk, Axially symmetric elastic waves in a laminated compressible composite material with initial stresses (Russian). Prikl.Mekh. 20 (1984), No. 7, 3-10.

[164] A. Harari, Generalized non-linear free vibrations of prestressed plates and shells. Int.J.Non-Linear Mech. 11 (1976), 169-181.

[165] M. Hayes and R.S. Rivlin, Propagation of a plane wave in an istropic elastic material subjected to pure homogeneous deformation. Arch. Rational Mech.Anal. 8 (1961), 15-22.

[166] M. Hayes and R.S. Rivlin, Surface waves in deformed elastic materials. Arch.Rational Mech.Anal. 8 (1961), 358-380.

[167] M. Hayes, Wave propagation and uniqueness in prestressed elastic solids. Proc.Roy.Soc.London Ser. A 274 (1963), 500-506.

[168] M. Hayes, Uniqueness for the mixed boundary value problem in the theory of small deformations superimposed upon large. Arch. Rational Mech.Anal. 16 (1964), 238-242.

[169] M. Hayes, On the displacement boundary-value problem in linear elastostatics. Quart.J.Mech.Appl.Math. 19 (1966), 151-155.

[170] M. Hayes and R.J. Knops, On the displacement boundary-value problem of linear elastodynamics. Q.Appl.Math. 26 (1968), 291-293.

[171] M.A. Hayes and R.S. Rivlin, Energy propagation in a deformed elastic material. Arch. Rational Mech.Anal. 45 (1972), 54-62.

[172] M.A. Hayes and R.S. Rivlin, Propagation of sinusoidal small-amplitude waves in a deformed viscoelastic solid. II. J. Acoust. Soc. Amer. 51 (1972), 1652-1663.

[173] M.A. Hayes and R.S. Rivlin, A class of waves in a deformed visco-elastic solid. Quart.Appl.Math. 30 (1972), 363-367.

[174] M. Hayes and C.O. Horgan, On the Dirichlet problem for incompressible elastic materials. J.Elasticity 4 (1974), 17-25.

[175] K. Häusler, On the viscoelastic properties of prestrained rubber. ZAMP 34 (1983), 25-50.

[176] G. Herrmann and A.E. Armenàkas, Vibrations and stability of plates under initial stress. J.Engng.Mech.Div., Proc.ASCE EM 3 (1960), 65-73.

[177] R. Hill, On uniqueness and stability in the theory of finite elastic strain. J.Mech.Phys.Solids 5 (1957), 229-241.

[178] R. Hill, Acceleration waves in solids. J.Mech.Phys.Solids 10 (1962), 1-16.

[179] J.M. Hill, Closed form solutions for small deformations superimposed upon the simultaneous inflation and extension of a cylindrical tube. J. Elasticity 6 (1976), 113-123.

[180] J.M. Hill, Self-adjoint differential equations arising in finite elasticity for small superimposed deformations. Int.J. Solids Structures 13 (1977), 813-822.

[181] I. Hlaváček, Variational formulation of the Cauchy problem for equations with operator coefficients. Aplik.Mat. 16 (1971), 46-63.

[182] L. van Hove, Sur l'extension de la condition de Legendre du calcul des variations aux integrales multiples à plusieurs fonctions inconnues. Proc.Kon.Ned.Akad.Wet. A 50 (1947), 18-23.

[183] R.K.T. Hsieh, Micropolarized and magnetized media. In Mechanics of Micropolar Media (Edited by O. Brulin and R.K.T. Hsieh), pp. 281-394, Singapore, World Scientific, 1982.

[184] D. Iannéce, A stability criterion for the equilibrium of a thermoelastic dielectric in the presence of conductors, Atti Accad. Naz. Lincei Rend. Cl.Sci.Fis.Mat.Natur. 68 (1980), 526-532.

[185] D. Ieşan, Principes variationnels dans la théorie de la thermoélasticité couplée. An.St.Univ."Al.I.Cuza" Iaşi, Sect.I, Matematica 12 (1966), 439-456.

[186] D. Ieşan, Sur la théorie de la thermoélasticité micropolaire couplée. C.R.Acad.Sci.Paris 265 A (1967), 271-275.

[187] D. Ieşan, Asupra deformarii plane generalizate. St.Cerc.Mat. 21 (1969), 595-609.

[188] D. Ieşan, On the linear coupled thermoelasticity with two temperatures. ZAMP 21 (1970), 583-591.

[189] D. Ieşan, On the thermal stresses in beams. J.Eng.Math. 6 (1972), 155-163.

[190] D.Ieşan, On some reciprocity theorems and variational theorems in linear dynamic theories of continuum mechanics. Memorie dell' Acad. Sci. Torino, Cl.Sci.Fis.Mat.Nat., Serie 4, n. 17, 1974.

[191] D. Ieşan, Thermal stresses in micropolar elastic cylinders. Acta Mechanica 21 (1975), 261-272.

[192] D. Ieşan, Saint-Venant's problem for inhomogeneous and anisotropic elastic bodies. J.Elasticity 6 (1976), 277-294.

[193] D. Ieşan, Theory of Thermoelasticity (in Romanian), Romanian Academy Publ. House, Bucharest, 1979.

[194] D. Ieşan, Thermal stresses in composite cylinders. J. Thermal Stresses 3 (1980), 495-508.

[195] D. Ieşan, Incremental equations in thermoelasticity. J. Thermal Stresses 3 (1980), 41-56.

[196] D. Ieşan, Thermoelastic stresses in initially stressed bodies with microstructure. J. Thermal Stresses 4 (1981), 387-405.

[197] D. Ieşan, On Saint-Venant's problem. Arch.Rational Mech.Anal. 91 (1986), 363-373.

[198] D. Ieşan, Saint-Venant's Problem. Lecture Notes in Mathematics. Vol. 1279, Springer-Verlag, Berlin, Heidelberg, New York, London, Paris, Tokyo, 1987.

[199] D. Ieşan, A theory of initially stressed thermoelastic materials with voids. An.St.Univ."Al.I.Cuza" Iaşi, s.Matematica, 33 (1987),167-184.

[200] D. Ieşan, Thermoelasticity of initially heated bodies. J. Thermal Stresses, 11 (1988), 17-38.

[201] J. Ignaczak, A completeness problem for stress equations of motion in the linear elasticity theory. Arch.Mech.Stosow. 15 (1963), 225-235.

[202] J. Ignaczak, Tensorial equations of motion for an elastic medium under initial stresses. In "New Problems in Mechanics of Continua" (Edited by O. Brulin and R.K.T. Hsieh), University of Waterloo Press, 1983, pp. 115-128.

[203] V. Ionescu-Cazimir, Problem of linear coupled thermoelasticity. Theorems on reciprocity for the dynamic problem of coupled thermoelasticity. I. Bull.Acad.Polon.Sci. Sér.Sci.Techn. 12 (1964), 473-480.

[204] V. Ionescu-Cazimir, Problem of linear coupled thermoelasticity. IV. Uniqueness theorem. Bull.Acad.Polon.Sci., Sér.Sci.Techn. 12 (1964), 565-573.

[205] S. Itou and A. Atsumi, The slipless indentation problem for an initially stressed neo-Hookean circular cylinder. Z.Angew.Math.Mech. 52 (1972), 243-245.

[206] T.P. Ivanov and A.I. Rachev, Thermoelastic waves in a prestressed circular cylinder. Theoret.Appl.Mech. 6 (1975), 53-63.

[207] D.D. Ivlev, On perfectly plastic flow of material with residual microstresses taken into account (Russian), Prikl.Mat.Mekh. 26 (1962), 709-714.

[208] M.G. Jakovenko, Influence of finite pre-deformations on an additional small loading of a non-linearly elastic half-plane (Russian). Tr.MVTU, 301 (1979), 72-80.

[209] F. John, Continuous dependence on data for solutions of partial differential equations with a prescribed bound. Comm. Pure Appl. Math. 13 (1960), 551-585.

[210] F. John, Hyperbolic and parabolic equations. Partial Differential Equations. (Edited by L. Bers and M. Schecter), pp. 1-123, New York, Wiley, 1964.

[211] N.F. Jordan and A.C. Eringen, On the static nonlinear theory of electromagnetic thermoelastic solids I, II. Int.J.Engng.Sci. 21 (1964), 59-115.

[212] Ju.I. Kadashevich and V.V. Novozhilov, Influence of initial micro-stresses on the macroscopic strain of polycrystals (Russian). Prikl.Mat.Meh. 32 (1968), 908-922.
[213] V.V. Kalinchuk and I.B. Poliakova, On the generation of waves in a prestressed layer (Russian). Prikl.Mat.Mekh. 44 (1980), 320-326.
[214] V.V. Kalinchuk and I.B. Poliakova, On the excitation of a pre-stressed cylinder (Russian). Prikl.Mat.Mekh. 45 (1981), 384-389.
[215] S. Kaliski and W. Nowacki, Thermal excitations in coupled fields. In Progress in Thermoelasticity (Edited by W. Nowacki), Warsaw, 1969.
[216] A. Kalnins and V. Biricikoglu, Theory of vibration of initially stressed shells. J.Acoust.Soc.Amer. 51 (1972), 1697-1704.
[217] B.K. Kar and A.K. Pal, On the possibility of Love wave propgation under initial shear stress. Proc.Indian Natl.Sci.Acad., Part A 51 (1985), 686-688.
[218] T. Kato, Linear evolution equations of hyperbolic type, II. J.Math. Soc. Japan, 25 (1973), 648-666.
[219] L.M. Keer and R. Ballarini, Smooth contact between a rigid indenter and an initially stressed orthotropic beam. AIAA J. 21 (1983), 1035-1042.
[220] O.D. Kellog, Foundations of Potential Theory. Springer-Verlag, New York, 1967.
[221] J. Kiusalaas, W. Jaunzemis and J.C. Conway, A strain-gradient theory for prestrained laminates. Int.J. Solids Structures 9 (1973), 1317-1330.
[222] R.J. Knops and L.E. Payne, Uniqueness in classical elastodynamics. Arch.Rational Mech.Anal. 27 (1968), 349-355.
[223] R.J. Knops and L.E. Payne, Stability in linear elasticity. Int.J. Solids Structures 4 (1968), 1233-1242.
[224] R.J. Knops and L.E. Payne, Stability of the traction boundary value problem in linear elastodynamics. Int.J.Engng.Sci. 6 (1968), 351-357.
[225] R.J. Knops and L.E. Payne, Continuous data dependence for the equations of classical elastodynamics. Proc. Cambridge Phil.Soc. 66 (1969), 481-491.
[226] R.J. Knops and L.E. Payne, On uniqueness and continuous dependence in dynamical problems of linear thermoelasticity. Int.J.Solids Structures 6 (1970), 1173-1184.

[227] R.J. Knops and L.E. Paynes, Uniqueness theorems in linear elasticity. Springer Tracts in Natural Philosophy, Vol. 19. Berlin-Heidelberg-New York, Springer, 1971.
[228] R.J. Knops and L.E. Payne, Growth estimates for solutions of evolutionary equations in Hilbert space with applications to elastodynamics. Arch.Rational Mech.Anal. 41 (1971), 363-398.
[229] R.J. Knops and E.W. Wilkes, Theory of Elastic Stability, pp. 125-302 of Flügge's Handbuch der Physik, vol. VI a/3 (Edited by C. Truesdell). Springer-Verlag, Berlin-Heidelberg-New York, 1973.
[230] R.J. Knops and L.E. Payne, Some uniqueness and continuous dependence theorems for nonlinear elastodynamics in exterior domains. Appl. Anal. 15 (1983), 33-51.
[231] R.J. Knops and B. Straughan, Continuous dependence theorems in the theory of linear elastic materials with microstructure. Int.J. Eng. Sci. 14 (1976), 555-565.
[232] V.P. Koshman, Dynamics of an incompressible half-plane with initial strains (Russian). Prikl.Mekh. 16 (1980), No. 9, 98-103.
[233] V.P. Koshman, Lamb's plane problem for a compressible half-space with initial stresses. (Russian). Prikl.Mekh. 16 (1980), No. 10, 94-100.
[234] S. Kosinski, Strong discontinuity plane waves in initially strained elastic isotropic medium. Mech.Teor.Stosow. 19 (1981), 545-562.
[235] K. Kowalczyk and M. Krasinski, The extensible elastica of a prestressed beam with arbitrary boundary conditions. Int.J.Mech.Sci. 25 (1983), 387-395.
[236] A.G. Kulikowskij and E.I. Sveshnikova, Investigation of the shock adiabat of quasitransverse shock waves in a prestressed elastic medium. (Russian). Prikl.Mat.Mekh. 46 (1982), 831-840.
[237] V.D. Kupradze, Potential methods in the theory of elasticity. Israel Program for Scientific Translations, Jerusalem, 1965.
[238] M. Kurashige, Circular crack problem for initially stressed neo-Hookean solid. ZAMM 49 (1969), 671-678.
[239] M. Lal and R.S. Sikarwar, Stress distributions in a circular ring under internal pressure. J.Math.Phys.Sci. 19 (1985), 461-473.
[240] W. Larecki, Influence of an initial static strain on a finite-amplitude longitudinal wave propagation. Bull.Acad.Polon.Sci., Sér. Sci.Techn. 28 (1980), 167-174.

[241] G. Lebon, Variational Principles in Thermomechanics. In "Recent Developments in Thermomechanics of Solids" (Edited by G. Lebon and P. Perzyna), Wien-New York, Springer-Verlag, 1980.

[242] P.C.Y. Lee, Y.S. Wang and X. Markenscoff, Elastic waves and vibrations in deformed crystal plates. Proc. 27-th. Ann. Freq. Cont. Symp. Fort Monmouth, N.J. 1973, p.1-15.

[243] S.G. Leknitskii, Theory of Elasticity of An Anisotropic Elastic Body, Holden-Day, Inc., San Francisco, 1963.

[244] E.E. Levi, Sulle equazioni lineari totalmente ellitiche alle derivate parziale. Rend.Circ.Mat.Palermo 24 (1907), 275-287.

[245] H.A. Levine, On a theorem of Knops and Payne in dynamical linear thermoelasticity. Arch. Rational Mech.Anal. 38 (1970), 290-307.

[246] G. Lianis, Small deformation superposed on an initial large deformation in viscoelastic bodies. Proc. 4-th.Int.Congr.Rheol. 1963, 2 (1965), 109-119.

[247] E.N.K. Liao and P.G. Kessel, On dynamic response of prestressed cylindrical shells. Green's tensor technique. Int.J. Solids Structures 9 (1973), 703-724.

[248] A.E.H. Love, A Treatise on the Mathematical Theory of Elasticity, Fourth Edition, Cambridge University Press, 1934.

[249] R.K. Mal and P.R. Sengupta, Magnetoelasticity under initial stress. I. General theory and two dimensional problems. Proc. Indian Nat. Sci.Acad., Part A 53 (1987), 144-164.

[250] K. Manivachakan, Plane strain problem of a punch indenting an initially stressed neo-Hookean anisotropic half-plane. Acta Mech. 45 (1982), 65-72.

[251] G.A. Maugin, Deformable dielectrics. I. Field equations for a dielectric made of several molecular species. Arch.Mech. 28 (1976), 679-692.

[252] A.G. Maugin, Infinitesimal discontinuities in initially stressed relativistic elastic solids. Commun. Math. Phys. 53 (1977), 233-256.

[253] M.F. McCarthy, Propagation of plane acceleration discontinuities in hyperelastic dielectrics. Int.J.Engng.Sci. 3 (1965), 603-623.

[254] M.F. McCarthy and W.A. Green, The growth of plane acceleration discontinuities propagating into a homogeneously deformed hyperelastic dielectric material in the presence of a magnetic field. Int.J.Engng. Sci. 4 (1966), 403-422.

[255] M.F. McCarthy, Wave propagation in nonlinear magneto-thermoelasticity. Proc. Vibration Problems 8 (1967), 337-349.

[256] M.F. McCarthy, Thermodynamic influences on the propagation of waves in electroelastic materials. Int.J.Engng.Sci. 11 (1973), 1301-1316.

[257] M.F. McCarthy, Singular surfaces and waves. In vol. II of the Continuum Physics (Edited by A.C. Eringen), New York, Academic Press, 1975.

[258] M.F. McCarthy and P.M. O'Leary, Thermodynamic influences on the propagation of electromagnetic shock waves. Proc. Roy. Irish Acad., Sect.A 75 (1975), 85-96.

[259] R.A. Meric, Optimal boundary tractions for solids with initial strains. J.Appl.Mech. 52 (1985), 363-367.

[260] S.G. Mikhlin, The Problem of the Minimum of a Quadratic Functional. Holden-Day, Inc., San Francisco, 1965.

[261] B.K. Min, H. Kolsky, A.C. Pipkin, Viscoelastic response to small deformations superposed on a large stretch. Int.J. Solids Structures 13 (1977), 771-781.

[262] R.D. Mindlin, Polarization gradient in elastic dielectrics. Int. J. Solids Structures 4 (1968), 637-643.

[263] M. Mişicu, Echilibrul mediilor continue cu deformari mari. Stud. Cercet.Mec.Metal. 4 (1953), 31-53.

[264] S.I. Moiseenko, Wave propagation in elastic clyinders with initial deformations (Russian). Izv.Sev.Kavk.Nauchn.Tsentr.Vyssh.Shk.Estestv. Nauki 1 (1982), 25-28.

[265] C.B. Morrey, Second order elliptic systems of differential equations. Ann. Math. Studies 33 (1954), 101-159.

[266] K. Mrithyumjaya Rao and B. Kesava Rao, Longitudinal vibrations of a prestressed thermoelastic transversely isotropic circular cylinder. Proc. Natl.Acad.Sci.India, Sect. A 46 (1976), 263-271.

[267] A.C. Murray, Uniqueness and continuous dependence for the equations of elastodynamics without strain energy function. Arch. Rational Mech.Anal. 47 (1972), 195-204.

[268] A.C. Murray and M.H. Protter, The asymptotic behaviour of solutions of second order systems of partial differential equations. J. Diff. Equations 13 (1973), 57-80.

[269] D.A. Musaev, Flexural waves in an incompressible non-circular cylinder with small initial deformations (Russian). Izv.Akad.Nauk. Az.SSR, Ser.Fiz.Tekhn.Mat.Nauk No. 4 (1984), 44-46.

[270] G.A. Nariboli and B.L. Juneja, Wave propagation in an initially stressed hypo-elastic medium. Int.J.Non-LinearMech. 6 (1971), 13-25.

[271] C.B. Navarro and R. Quintanilla, On existence and uniqueness in incremental thermoelasticity, ZAMP 35 (1984), 206-215.

[272] R.E. Nickell and J.L. Sackman, Variational principles for linear coupled thermoelasticity. Quart.Appl.Math. 26 (1968), 11-26.

[273] J.P. Nowacki, Steady-state problems of thermopiezoelectricity. J.Thermal Stresses, 5 (1982), 183-194.

[274] W. Nowacki, Mathematical models of phenomenological piezo-electricity. In New Problems in Mechanics of Continua (Edited by O. Brulin and R.K.T. Hsieh), University of Waterloo Press, 1983, 29-50.

[275] J.L. Nowinski, Thermal waves in an elastic highly stretched cylindrical bar. Acta Mechanica, 7 (1969), 45-57.

[276] O. Nurzhumaev, Propagation of Love waves in an initially stressed medium with material boundary. (Russian). Izv.Akad.Nauk.Kaz.SSR, Ser.Fiz.-Mat. 3 (1980), 78-79.

[277] O. Nurzhumaev and Z. Karataev, Propagation of transverse waves in a layered medium involving initial stresses. (Russian). Izv. Akad. Nauk.Kaz.SSR, Ser.Fiz.Mat. 1 (110) (1983), 77-80.

[278] O. Nurzhumaev and Z. Karataev, Propagation of Love waves in a pre-stressed sandy medium with material boundary. (Russian). Izv. Akad. Nauk.Kaz.SSR, Ser.Fiz.Mat. 3 (1985), 63-65.

[279] R.W. Ogden, Waves in isotropic elastic materials of Hadamard, Green, or harmonic type. J.Mech.Phys.Solids 18 (1970), 149-163.

[280] Z. Olesiak and I.N. Sneddon, The distribution of thermal stress in an infinite elastic solid containing a penny-shaped crack. Arch. Rational Mech. Anal. 3 (1960), 238-254.

[281] D.R. Owen, J.A. Figueiras and F. Damjanic, Finite element analysis of reinforced and prestressed concrete structures including thermal loading. Comput. Methods Appl.Mech.Eng. 41 (1983), 323-366.

[282] U.C. Pan, SH-waves in two layered inhomogeneous medium under initial stress. Bull. Calcutta Math. Soc. 71 (1980), 86-93.

[283] U.C. Pan and S.K. Chakraborty, Love waves under initial stress. Math. Stud. 43 (1982), 301-304 (1975).
[284] M.R. Parameswara, K.R. Mrithyumjaya and B. Kesava Rao, Longitudinal vibrations of a prestressed, thermoelastic circular cylinder. Proc. Nat.Acad.Sci.India, Sect. A 43 (1973), 219-228.
[285] G. Paria, Magnetoelasticity and magnetothermoelasticity. In Advances in Applied Mechanics (Edited by G. Kuerti), Academic Press, New York, 1967.
[286] G. Paria, Cauchy theory of initial stress and its application to depression of semi-infinite elastic medium. Indian J. Engin. Math. 1 (1968), 121-126.
[287] G. Paria, Cauchy's structure theory of elastic solids with initial stresses. Indian J. Engin. Math. 2 (1969), 7-16.
[288] H. Parkus, Magneto-thermoelasticity. CISM Lecture Notes 118, Springer-Verlag, Berlin,1972.
[289] A. Pazy, Semigroups of Linear Operators and Applications to Partial Differential Equations. Springer, New York, Berlin, Heidelberg, Tokyo, 1983.
[290] C.E. Pearson, General theory of elastic stability. Quart. Appl. Math. 14 (1956), 133-144.
[291] W. Pietraszkiewicz, Stress in isotropic elastic solid under superposed deformations. Arch. of Mech. 26 (1974), 871-884.
[292] A.C. Pipkin and R.S. Rivlin, The formulation of constitutive equations in continuum physics. Div.Appl.Math., Brown University Report, September, 1958.
[293] A.C. Pipkin and R.S. Rivlin, Electrical conduction in a stretched and twisted tube. J.Math.Phys. 2 (1961), 636-638.
[294] A.C. Pipkin and R.S. Rivlin, Small deformations superposed on large deformations in materials with fading memory. Arch. Rational Mech. Anal. 8 (1961), 297-308.
[295] P.Podio-Guidugli, St.Venant formulae for generalized St. Venant problems. Arch.Rational Mech.Anal. 81 (1983), 13-20.
[296] V.P. Poroshin, Penetration of a stamp into a prestressed physically nonliner elastic layer. (Russian). Izv.Akad.Nauk Arm.SSR, Mekh. 39 (1986), No. 2, 24-30.

[297] W. Prager, The general variational principle of the theory of structural stability. Quart. Appl. Math. 4 (1946), 378-384.

[298] B. Prasad, On the response of a Timoshenko beam under initial stress to a moving load. Int.J.Engng.Sci. 19 (1981), 615-628.

[299] A. Prechtl, Deformable bodies with electric and magnetic quadrupoles. Int.J.Engng.Sci. 18 (1980), 665-680.

[300] R. Quintanilla, A short note on incremental thermoelasticity. Stochastica 7 (1983), 163-167.

[301] R. Quintanilla and H.T. Williams, An existence and uniqueness theorem for incremental viscoelasticity. Q.Appl.Math. 43 (1985), 287-294.

[302] S. Rajagopal, The role of initial stress in lattice dynamics. Ann. Phys. 6 (1960), 182-191.

[303] J. Ramakanth, Some problems of propagation of waves in prestressed isotropic bodies. Proc. Vibration Problems 6 (1965), 161-172.

[304] J. Ramakanth, Radial vibrations of a prestressed sphere. Bull. Akad. Polon.Sci.Sér.Sci.Techn. 13 (1965), 401-408.

[305] C.V. Raman and K.S. Viswanathan, On the theory of elasticity of crystals. Proc.Ind.Acad.Sci., Ser.A 42 (1965), 51-61.

[306] F. Rammerstorfer, On the optimal distribution of the Young's modulus of a vibrating prestressed beam. J. Sound and Vibr. 37 (1974), 140-145.

[307] F.G. Rammerstorfer, Increase of the first natural frequency and buckling load of plates by optimal fields of initial stresses. Acta Mechanica 27 (1977), 217-238.

[308] D.P. Reddy and J.D. Achenbach, Simple waves and shock waves in a thin prestressed elastic rod. ZAMP 19 (1968), 473-485.

[309] H. Reismann and H.H. Liu, Forced motion of an initially stressed rectangular plate - an elasticity solution. J. Sound Vibration 55 (1977), 405-418.

[310] H. Reismann and Z.A. Tendorf, Dynamics of initially stressed plates. J. Appl. Mech. 43 (1976), 304-308.

[311] H. Reismann and P.S. Pawlik, Dynamics of initially stressed hyper-elastic solids. ZAMM 59 (1979), 145-155.

[312] S. Rionero and G.P. Galdi, The weight function approach to uniqueness of viscous flows in unbounded domains. Arch.Rational Mech.Anal 69 (1969), 36-52.

[313] S. Rionero and S. Chiriţa, The Lagrange identity method in linear thermoelasticity. Int.J.Engng.Sci. 25 (1987), 935-947.

[314] R.S. Rivlin, Large elastic deformations of isotropic materials. II. Some uniqueness theorems for pure, homogeneous deformations. Phil. Trans.Roy.Soc.London, Ser.A 240 (1948), 491-508.

[315] R.S. Rivlin, Foundations of Continuum Thermodynamics (Edited by N.Nina and J.Whitelaw), McMillan, London, 1975.

[316] C. Rogers, T. Moodie, T. Bryant and D.L. Clements, Radial propagation of rotary shear waves in an initially stressed neo-Hookean material. J. Mec. 15 (1976), 595-614.

[317] E. Rusu, Small thermoelastic deformation superposed on finite thermoelastic deformation in a cylindrical tube. Bull. Inst. Politehnic din Iaşi, Sect. Mecanica, 25 (1979), 19-26.

[318] K.N. Sawyers and R.S. Rivlin, Instability of an elastic material. Int.J. Solids Structures 9 (1973), 607-611.

[319] K.N. Sawyers and R.S. Rivlin, On the speed of propagation of waves in a deformed elastic material. ZAMP 28 (1977), 1045-1057.

[320] K.N. Sawyers and R.S. Rivlin, On the speed of propagation of waves in a deformed compressible elastic material. ZAMP 29 (1978), 245-251.

[321] P. Seide, Large deflection of prestressed simply supported rectangular plates under uniform pressure. Int.J. Non-Linear Mech. 13 (1978), 145-156.

[322] I.K. Senchenkov, Extremal principles of the theory of small thermoviscoelastic deformations superposed on fully established finite deformations. (Russian). Prikl.Mekh. 16 (1980), No. 7, 26-34.

[323] R.T. Shield and R.L. Fosdick, Extremum principles in the theory of small elastic deformations superposed on large elastic deformations. Progress in Applied Mechanics (Prager Anniversary Volume), Macmillan, New York, 1963, 107-125.

[324] R.T. Shield, On the stability of linear continuous systems. ZAMP 16 (1965), 649-686.

[325] R.T. Shield, An energy method for certain second-order effects with application to torsion of elastic bars under tension. J. Appl.Mech. 47 (1980), 75-81.

[326] R.T. Shield and S. Im, Small strain deformations of elastic beams and rods including large deflections. ZAMP 37 (1986), 491-513.

[327] R.S. Sidhu and S.J. Singh, Reflection of P and SV waves at the free surface of a prestressed elastic half-space. J.Acoust.Soc. Am. 76 (1984), 594-598.

[328] R.L. Sierakowski, C.T. Sun and I.K. Ebcioglu, Instability of a hollow-rubber-like cylinder under initial stress. Int.J.Non-Linear Mech. 10 (1975), 193-205.

[329] M. Slemrod and E.F. Infante, An invariance principle for dynamical systems on Banach space: Application to the general problem of thermo-elastic stability. IUTAM Symposium on Instability of Continuous Systems-Hersenalb., Springer-Verlag, New York, 1969.

[330] A.M. Slobodkin, On the justification of the principle of minimum potential energy in problems of equilibrium stability of nonlinearly elastic membranes. Prinkl.Mat.Meh. 35 (1971), 168-171.

[331] E. Soós, Discrete and Continuous Models of Solids (Romanian), Editura Stiinţifica, Bucureşti, 1974.

[332] E. Sternberg and J.K. Knowles, Minimum energy characterizations of Saint-Venant's solution to the relaxed Saint-Venant problem. Arch. Rational Mech.Anal. 21 (1966), 89-107.

[333] C.T. Sun and B.V. Sankar, Smooth indentation of an initially stressed orthotropic beam. Int.J.Solids Structures 21 (1985), 161-176.

[334] E.I. Sveshnikova, Quasi-transverse shock waves in an elastic medium in the case of special types of initial strain. (Russian). Prikl. Mat. Mekh. 47 (1983), 673-678.

[335] E.S. Şuhubi, Small torsional oscillations of a circular cylinder with finite electric conductivity in a constant axial magnetic field. Int.J.Engng.Sci. 2 (1964), 441-460.

[336] E.S. Şuhubi, Small longitudinal vibration of an initially stretched circular cylinder. Int.J.Engng.Sci. 2 (1965), 509-517.

[337] E.S. Şuhubi, The growth of acceleration waves of arbitrary form in deformed hyperelastic materials. Int.J.Engng.Sci. 8 (1970), 699-710.

[338] E.S. Şuhubi, Propagation of a plane wave in an initially stressed thermoelastic medium. Bull.Techn.Univ. Istanbul, 24 (1972), 54-73.

[339] B. Tabarrok and S. Dost, Application of stress functions to initial stress problems. J.Appl.Mech. 45 (1978), 350-354.

[340] R.J. Tait, J.B. Haddow and T.B. Moodie, A note on infinitesimal shear waves in a finitely deformed elastic solid. Int.J.Eng.Sci. 22 (1984), 823-827.

[341] S. Tan, Wave propagation in initially stressed elastic solids. Acta Mechanica, 4 (1967), 92-106.
[342] C. Teodosiu, Nonlinear theory of the materials of grade two with initial stresses and hyperstress. II. Constitutive equations. Bull. Acad.Polon.Sci.Ser.Sci.Techn. 15 (1967), 95-102.
[343] T.Y. Thomas, Extended compatibility conditions for the study of surfaces of discontinuity in continuum mechanics. J.Math.Mech. 6 (1957), 311-322.
[344] R.N. Thurston and K.Brugger, Third-order elastic constants and the velocity of small amplitude waves in homogeneously stressed media. Phys.Rev. 133 (1964) A 1604 - A 1610.
[345] R.N. Thurston, Effective elastic coefficients for wave propagation in crystals under stress. J. Acoust. Soc. Amer. 37 (1965), 348-356.
[346] H.F. Tiersten, On the nonlinear equations of thermoelectroelasticity. Int.J.Engng.Sci. 9 (1971), 587-593.
[347] S. Timoshenko and J.N. Goodier, Theory of Elasticity, 2nd ed., McGraw Hill, New York, 1951.
[348] T. Tokuoka and Y. Iwashimiza, Acoustical birefringence of ultrasonic waves in deformed isotropic elastic materials. Int. J. Solids Structures 4 (1968), 383-389.
[349] T. Tokuoka and M. Saito, Elastic wave propagations and acoustical birefringence in stressed crystals. J.Acoust.Soc.Amer. 45 (1969), 1241-1246.
[350] R.A. Toupin and B. Bernstein, Sound waves in deformed perfectly elastic materials. Acoustoelastic effect. J. Acoust.Soc. Am. 33 (1961), 216-225.
[351] R.A. Toupin, A dynamical theory of elastic dielectrics. Int. J. Engng. Sci. 1 (1963), 101-126.
[352] A.R. Toupin and D.C. Gazis, Surface effects and initial stress in continuum and lattice models of elastic crystals. Lattice Dynamics (Edited by R.F. Wallis), New York, Pergamon Press, 1965.
[353] C. Truesdell, The rational mechanics of materials - past, present, future. Appl.Mech. Reviews 12 (1959), 75-80.
[354] C. Truesdell and R. Toupin, The classical field theories. In vol. III/1 of the Handbuch der Physik (Edited by S. Flügge), Berlin-Heidelberg-New York, Springer-Verlag, 1960.

[355] C. Truesdell, General and exact theory of waves in finite elastic strain. Arch. Rational Mech. Anal. 8 (1961), 263-296.

[356] C. Truesdell and W. Noll, The non-linear field theories of mechanics. In vol. III/3 of the Handbuch der Physik (Edited by S. Flügge), Berlin-Heidelberg-New York, Springer-Verlag, 1965.

[357] C. Truesdell, Existence of longitudinal waves. J. Acoust. Soc. Am. 40 (1966), 729-730.

[358] C. Truesdell, The rational mechanics of materials - past, present, future (Corrected and modified reprint of [353]), pp. 225-236, of Applied Mechanics Surveys, Spartan Books, 1966.

[359] C. Truesdell, History of Classical Mechanics. Die Naturwissenchaften 63, Part I: to 1800, 53-62; Part II: the 19-th and 29-th Centuries, 119-130. Springer-Verlag, 1976.

[360] C. Truesdell, Some challenges offered to analysis by rational thermo-mechanics, pp. 495-603 of Contemporary Developments in Continuum Mechanics and Partial Differential Equations (Edited by G.M. de la Penha and L.A. Medeiros), North-Holland, 1978.

[361] C. Truesdell, Rational Thermodynamics. New York, McGraw-Hill, 1969.

[362] W. Urbanowski, Small deformations superposed on finite deformations of a curvilinearly orthotropic body. Arch. Mech. Stosow. 11 (1959), 223-241.

[363] W. Urbanowski, Deformed body structure. Arch. Mech. Stosow. 13 (1961), 277-294.

[364] P.D.S. Verma and H.R. Chaudhry, Small deformation superposed on large deformation of an elastic dielectric. Int. J.Engng.Sci. 4 (1966), 235-247.

[365] D.K. Wagh, Effect of initial stress on the distribution of pressure producing a crack of prescribed shape. J. Indian Acad. Math. 7 (1985), 12-18.

[366] D.C. Wallace, Thermoelasticity of stressed materials and comparison of various elastic constants. Phys. Review, II. Ser. 162 (1967) 776-789.

[367] K. Walton, Seismic waves in pre-strained media. Geophys. J. Roy. Astron.Soc. 31 (1973), 373-394.

[368] J.H. Weiner, A uniqueness theorem for the coupled thermoelastic problem. Quart. Appl.Math. 15 (1957), 102-105.

[369] Z. Wesolowski, Stability in some cases of tension in the light of the theory of finite strain. Arch.Mech.Stosow. 14 (1962), 875-900.

[370] Z. Wesolowski, The axially symmetric problem of stability loss of an elastic bar subject to tension. Arch. Mech. Stosow. 15 (1963), 383-395.

[371] Z. Wesolowski, Acoustic wave in finitely deformed elastic material. Arch. of Mech. 24 (1972), 793-801.

[372] Z. Wesolowski, Small vibrations of elastic medium deforming in time. Arch. of Mech. 24 (1972), 573-586.

[373] Z. Wesolowski and J. Maczynski, Stability of an initially prestressed elastic strip. Rozprawy inz. 23 (1975), 687-696.

[374] Z. Wesolowski, Strong discontinuity wave in initially strained elastic medium. Arch.Mech. 30 (1978), 309-322.

[375] Z. Wesolowski, Wave propagation in finitely deformed elastic material. Nonlinear dynamics of elastic bodies. CISM Courses Lect. 227, 81-142, 1978.

[376] A.B. Whitman, Constitutive equations for curved and twisted, initially stressed elastic rods. Acta Mechanica 30 (1978), 237-257.

[377] E.W. Wilkes, On the stability of a circular tube under end thrust. Quart.J.Mech.Appl.Math. 8 (1955), 88-100.

[378] N.S. Wilkes, Continuous dependence and instability in linear thermo-elasticity. SIAM J. Math. Anal. 11 (1980), 292-299.

[379] R.A. Williams and L.E. Malvern, Harmonic dispersion analysis of incremental waves in uniaxially prestressed plastic and viscoplastic bars, plates, and unbounded media. J.Appl.Mech. 36 (1969), 59-64.

[380] A.J. Willson, The propagation of magneto-thermo-elastic plane waves. Proc. Camb. Phil.Soc. 59 (1963), 483-490.

[381] A.J. Willson, Surface and plate waves in biaxially-stressed elastic media. Pure and Appl. Geophys. 102 (1973), 182-192.

[382] A.J. Willson, Wave propagation in thin prestressed elastic plates, Int.J.Engng.Sci. 15 (1977), 245-251.

[383] A.J. Willson, Plate waves in Hadamard materials. J. Elasticity 7 (1977), 103-111.

[384] T. Woo and R.T. Shield, Fundamental solutions for small deformations superposed on finite biaxial extension of an elastic body. Arch. Rational Mech.Anal. 9 (1962), 196-224.

[385] K. Yosida, Functional Analysis. Springer-Verlag, Berlin-Heidelberg-New York, 1968.

[386] C.P. Yu and S. Tang, Magneto-elastic waves in initially stressed conductors. ZAMP 17 (1966), 766-775.

[387] S. Zahorski, Small additional motion superposed on the fundamental motion of a hypoelastic medium. Bull. Acad. Polon. Sci. Sér.Sci. Techn. 11 (1963), 449-454.

[388] O.P. Zhuk, Waves on the interface of prestressed elastic bodies. (Ukrainian), Dopov.Akad.Nauk.Ukr.SSR, Ser. A (1979), 818-820.

[389] H. Zorski, On the equations describing small deformation superposed on finite deformation. Proc. International Symposium Second-Order Effects, Haifa, 1962, Macmillan, New York, 1964, pp. 109-128.

[390] L.M. Zubov, Variational principles of the nonlinear theory of elasticity. Case of superposition of a small deformation on a finite deformation. (Russian). Prikl.Mat.Mekh. 35 (1971), 848-852.

[391] L.M. Zubov, Theory of small deformations of prestressed thin shells. J. Appl. Math.Mech. 40 (1976), 73-82.

[392] L.M. Zubov and A.N. Rudev, Homogeneous solutions for a prestessed elastic plate. (Russian). Prikl.Mat.Mekh. 42 (1978), 920-929.

Index